# The Bricks -> Scissors of Creation of Life and matter or matter m1 = process (room) p...

Our matter and life creating from the smallest functional bricks (matter = process -> matter... not exist ) Eskm1DAA <- EvmDNA.

# Introduction

Independent futuristic writer. Project about new concepts of use economically, technological tools for reshaping the environment for ecologic and development goals by space explorations and medical matters. Books " Masters of life and universe", "100 verses for life and space re-creation", "Neo Creation of Life and Space/Matter or Bricks of Micro Drone Revolution", " Towards the foundation of the mechanism of neo-creation", "Beyond the nature of life and stars", " Neo evolutionary blessing", " A handbook of re-neo processing of life and matter". Studied pedagogic, technology, economy, administration, law, and politics at various academies and universities.

All states of the environment have a process basis; they are a room for processes or rather processes for these rooms of matter and life.

Moving aside the hitherto classical physical measures such as the earth, the sun, time, mass, energy, matter ... only dynamic process relations at a given level of perception of the so-called process rooms - their perception, experience, and creation.
View of relations between processes, inter-process, inter-process, process-related
mass, light, time -primitive readout may suffice on some degenerative level -centrically, but process potential process systems ..alternative responsibility for
covered, used, to be used, covered process systems
This is not about contrariness, change for change, academic discussions (although this hyperevolutionary tone is intended to stimulate proper/more responsible for literally co-shaping the environment) economic, scientific, social, ecological, political relations), but creating in educational (literally -> AI) of the next generation of tools to change modern process systems in any direction of the system, whether our external or internal ... because these classic centers no longer matter, at least at a different technological, cognitive, useful/applicable/valid level - perhaps reaching other references, centers, factors of our development as the basis of existence, life, and matter.

So we are talking about a dynamic shift of these centric ones taking place before our eyes, which will become only a singularity background, or maybe just the beginning of the subsequent explosion of development - the opposite of degeneration - within the framework I suggest - singularization/inflation at a given stage of development, niche .... production technology economy neostrategy, neoevolution, neophysics, neobiology, neoreligion...

Where the previous efficiency, structure, and organization, EvmDNA processing is recognized, developed as a current/old material, source, raw material, component - m1 needed for further improvement of the organization, processing -> Eskm1DAA environment, which is to be the following m2 material, at further stages of this process $p = EvmDNA = m1 < Eskm1DAA = m2 < Eskm2DAA$, etc., where Ev is environmentally indifferent vanity economics, current efficiency, m - infrastructure, environment, mass, matter, DNA - as the current broadly understood control and administration system, Esk as selective economics efficiency - separately worked out, consciously, and the DAA is, in short, the future dynamic active administration.

I will find broader interdisciplinary applications and arguments not so easy to read and study in this book.

It is not fine literature, but a literature of change, a new unframed language, because the new unframed, non-geo-egocentric proposed structure of change, functioning in a dynamic environment of dynamic processes, co-participating, competing systems of this environment, which we also involuntarily shape.

Singularization of old centers, planetary, virtual, at some level even useful, existing factors. Abandonment, exit from old systems. Replacing them with more useful, more precise instrumental ones, literally piercing through and through the existing systems, models, systems, time centers, lights, masses, but DAA _> processroom of these changes.

Process structure = matter
re-neo matter always because it is hidden behind the mask/face of the matter.
There are processes with infinite dynamic connections, dynamic economic ones, in which we can participate passively or more or less consciously.
You can't do it right away, but there is always a more or less winding road, aware of the dematerialization of the image's fossils.

Its speed, the agility of realization depends on the contribution of the economics of the production of the investment environment to this process; simply, in other words … organization by the organization of science -> economics of environmental infrastructure = economy = responsibility on an infinite scale of an endless environment of processes, processrooms.

Science is supposed to streamline this process, and in turn, the infrastructure economics of this environment is considered to support this science.

The weighted average of this intentionality, investment awareness, is the proper property result of a given phenomenon, object, process matter at every level, the direction of activity known to us, less known or unknown to systems->processes.

Redirection of values, the size of the Big Bang - this is already done in artificially produced micro suns, micro Big Bangs
crazy?
It is madness to stand in the place of the contemporary suffocating environment and our inertia of behavior, often expressed in economic, scientific, schizophrenic, geo-centric, ego-centric, always fatalistic, ignorant, entropic models.

That's why open structure, expediency of EskmDAA as a room process is so important, not the next stage, the material level for further processing, but m as p, as always process room stimulation, control, life strategy.

It doesn't matter his, this m, and already in the new approach p, the further internal arrangement of this matter m=p process, or rather just p as processroom
All classical coefficients, physical processes, including time, energy, mass, are subject to process

singularization and ... economic evolutionary, including, of course, entropic ... on our part - the question is what direction, what next steps will be taken to always be above this process dynamic transformation of the infrastructure environment, how to be less singularity trash, to be more in the vanguard of change, development -> existence, i.e., life, not degradation. This ever-advancing wall of singularity is dependent on the environment as well as us, who would be the accomplices that we have always been, are, and will be; only a matter of whether more active, responsible participants in this process.

All states of the environment have a process basis; they are a room for processes or rather processes for these rooms of matter and life, the matter of life.

Each element, quark, has its processable root $EskmDAA = m1,2...n \to p$ determining its properties at some level, the utility, which is open against the claustrophobic/centric vision of matter and life.

Here, m...n has a secondary character - it's about prime factors, not m, but p as a whole from the EskmDAA room, as a set of m that is directed, transformed, stimulated, auto-adjusted, autonomic, dynamic, structural p

p1> Ev/EskmDNA/DAA <pn process superset ultimately in the perception of the process of a given level m as the actual envelope of a given object, phenomenon, undertaking...

In his time, the famous Italian director Antonioni in the film "Bolw up", pointed out that reality does not exist because before it is defined, it disappears in ... in this dynamic world.

...So we need to be more flexible, process-oriented, systematically take into account the changing infrastructure of the environment, sometimes it can simply overtake it ... put the matter before the fait accompli directing, substituting, supplementing in an ongoing process, being that active aware, direction, horizontal, imposing them

EskmDAA=p no longer m...n but dynamic p against static

mn = EskmDAA/EvmDNA in an infinite sea of processrooms where we are shareholders. It's a question of responsibility, that is, competence, a contribution to the properties of the microcosm of matter and life.

Efficiency, productivity, autonomy concerning a given element of the user - the economics of macro-micro space ->neo-neolithic revolution in the production of space...

The rest of the introduction showed quite truthfully in the table of contents.

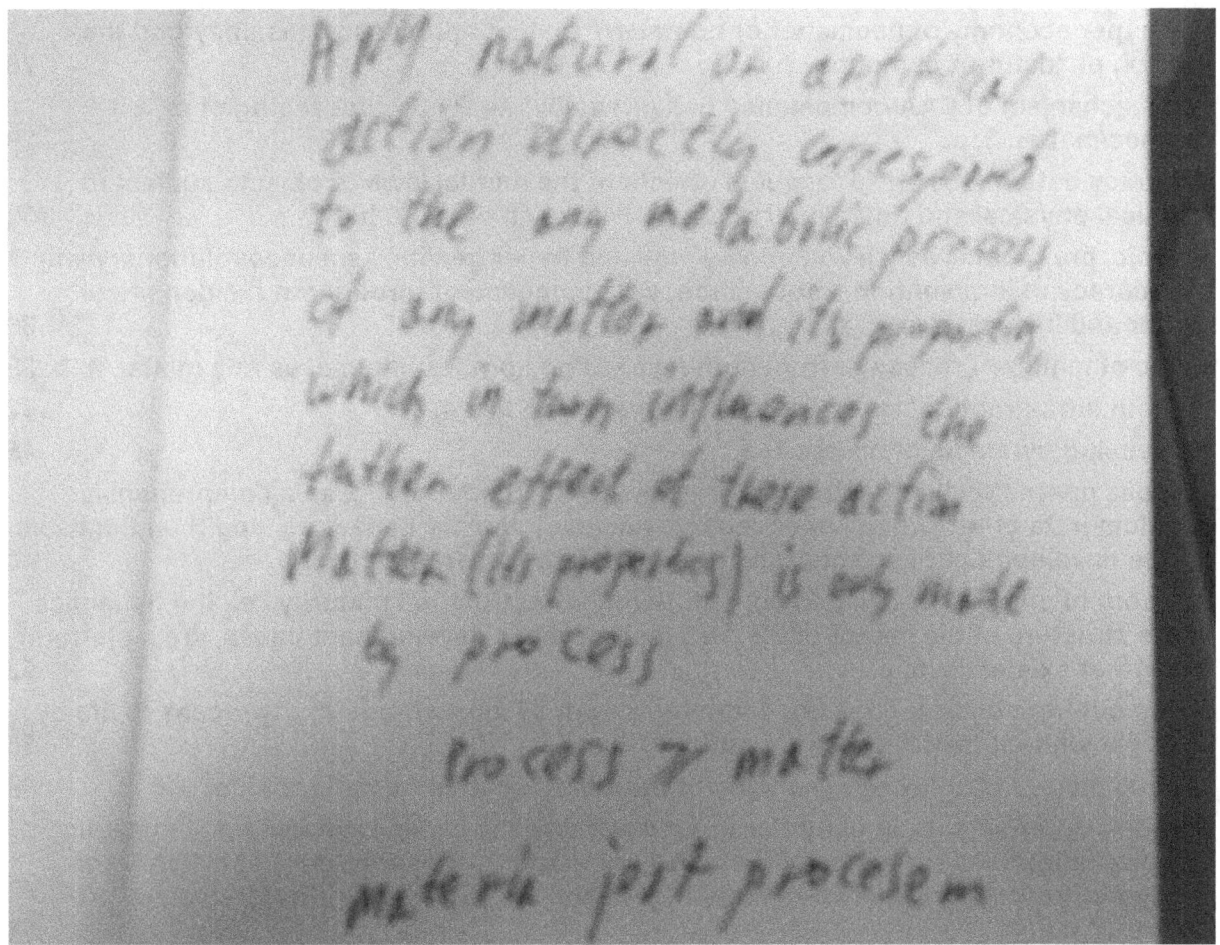

Towards the application of the mechanism – the infinite fuel potential of the economic explosion of each selected segment of the process – the process room, which is the basis for the derivatives of material-centric values, including cells, elementary particles... 1. a.

# Towards the application of the mechanism – the infinite fuel potential of the economic explosion of each selected segment of the process – the process room, which is the basis for the derivatives of material-centric values, including cells, elementary particles... 1. a.

Eskm1DAA = P – processing, specifying, directing the flow of ... the infinite cosmic ocean of micro-macro processors ... micro macro processors in the standardization of development control, i.e., existence, where m1 = EvmDNA of the selected period, a segment of the phenomenon, object of matter, life.

The proposed horizon of economic and environmental activities is open to all undertakings, actions more or less abstract, active, not passive, futuristic, existential, scientific, medical, economic, whose common denominator is the economy of processing processes –
=Eskm2DAA> m2; m2=Eskm1DAA > m1; m1 =EvmDNA,...
The proposed description of the world mechanism based on the processing of the infinite segmented reneokreative evolutionary potential of all phenomena supported by

further and derivative processes, where our perception tries to fortify them with some perceptually received image, models, mask, face, material casing.

The material description of the world of the process of process of …the process is ok. Still, it cannot overshadow the infinite potential behind this wall, which does not always sufficiently reflect the deeper sources, causes, and at the same time, endless possibilities that elude the attempt of a static, i.e., material description and influencing the environment in us and us outside.

Here, to aid the dynamic description of phenomena, hyperactive, not passive, but gigantic – because only such actions and voices count in nature and before God – influencing phenomena comes to a macro-economic description indicating the general directions, costs, development, and existence.

On the one hand, Esk > Ev, i.e., increasing taxes/platform -> and in the direction, in the final intention, increasing expenses indefinitely, trillion times … – unconditional increase in investment and, at the same time, competence, environmental and infrastructural responsibility on a not only global scale but micro and macrocosmic, and on the other hand, DAA > DNA, i.e., reducing the costs / mass production, investment – credit – interest rates, i.e., switching to cost-cutting activities as the primary mechanism of development from … nothing, i.e., redirecting sources, raw materials, semi-finished products, components as fuel for further development as the basis for any existence, repair, development of what we want to repair, develop at the level of physics, medicine, etc.

Giving more extraordinary investment grounds, not smothering the material potentials of the existing EvmDNA in the current "egocentric = geocentric" physical, biological descriptions ... but instead Esk as bank interest, investment finally condemning to infinite potential and demand, under active DAA as investment taxes, artificial addition of efficiency to a given state of EvmDNA, the same can be proposed in ... physics, biology ... energy, medicine, ... let's try to follow this thread.

This proposed mechanism is stimulation, ignition
of the development, the explosion of the enterprise of a cell, an atom, a plant...

It gives the freedom to develop the economy in the classical sense and backward science, enhancing the geometrical and algorithmic progress for off-scale development.

Not science without freedom, not only in words but based on building the foundations for its development.
It is giving hope for everyone, every initiative, opening the borders figuratively and literally (they lose their meaning with such an enormous pace of development) of a new BigBang

Demonstration of freedom and development Eskm(1...n)DAA...n
towards the mechanism
the explosion of the economic development of the cell and other process rooms, as the
appropriate pace of competence growth with competitive, also infinite process systems,
i.e., process rooms of nature, the environment.
segment forces of growth of selected process rooms from a given more or less
homogeneous system
As part of selective economics, lending to a given direction of efficiencies of a given
segment of a given economic project and investment tax in this above-described
segment

Subdue the earth, i.e., total responsibility, total competence, i.e., over matter, according
to God, because it expresses to us the meaning of existence, as well as the existence
of the world if we do not want it to be ground by other forces, systems, also based on
processes with which we can instead of destructive, we can establish cooperation on a
scale, or somewhat above the cosmic scale – because we have no obligation or rights
to recognize any material scale.

The wind will not overthrow these foundations of the proposed development because
the root of it is the above-described, literally engraved responsibility and... mercy for
oneself and others above the primitive pseudo-libertarian material pseudo, egocentric,
and simultaneous fatalistic material laws.

So the mercy of process responsibility over the so-called static EvmDNA can be found
in most patterns, models of modern science, technology...
And new foundations based on responsibility and, thus, total competence.
Let the spirit of renewal, a rebirth of the face of this earth enter, that is, let us try to
understand ourselves and our surroundings more responsibly – the proposed model,
language, spirit Eskm(1...n)DAA = Pn

Where is it ... in the constantly rebuilt, renewed competence of responsibility, dignity,
and mercy above the current economy/science of vanity Ev, because Esk has this cross
of responsibility to bear, understanding history and shaping the future. Whatever its
religious name, natural description

It is further learning, the open book of science, dignity, rights of the citizen, nation, and
the world from the microparticle upwards in our share ... in the formula Eskm1DAA = P...

Because despite the classic economic and scientific calculations .... backsight
the economy, the nature of the service is growing – this is an effect beyond m1 in the
so- called external factors but internal process factors as if at a distance controlling the
existing m1 system in terms of "external" Esk and DAA

Constantly adding this infinite Esk-DAA fuel to m1...
It does not reduce the m1's potential, but the issue of proper targeting actually increases it.

To be continued, the description of the mechanism of the explosion of development or the blast of degradation – whether we will be on one side or the other of the process of singularization – depends on the active or passive participation in these processes engraved above.

Is it possible to replace the parameters/laws of the macrocosm at deeper and further levels of the structure of matter, or rather its process causes? At some point - literally- Yes, and there are countless points. 1. b.

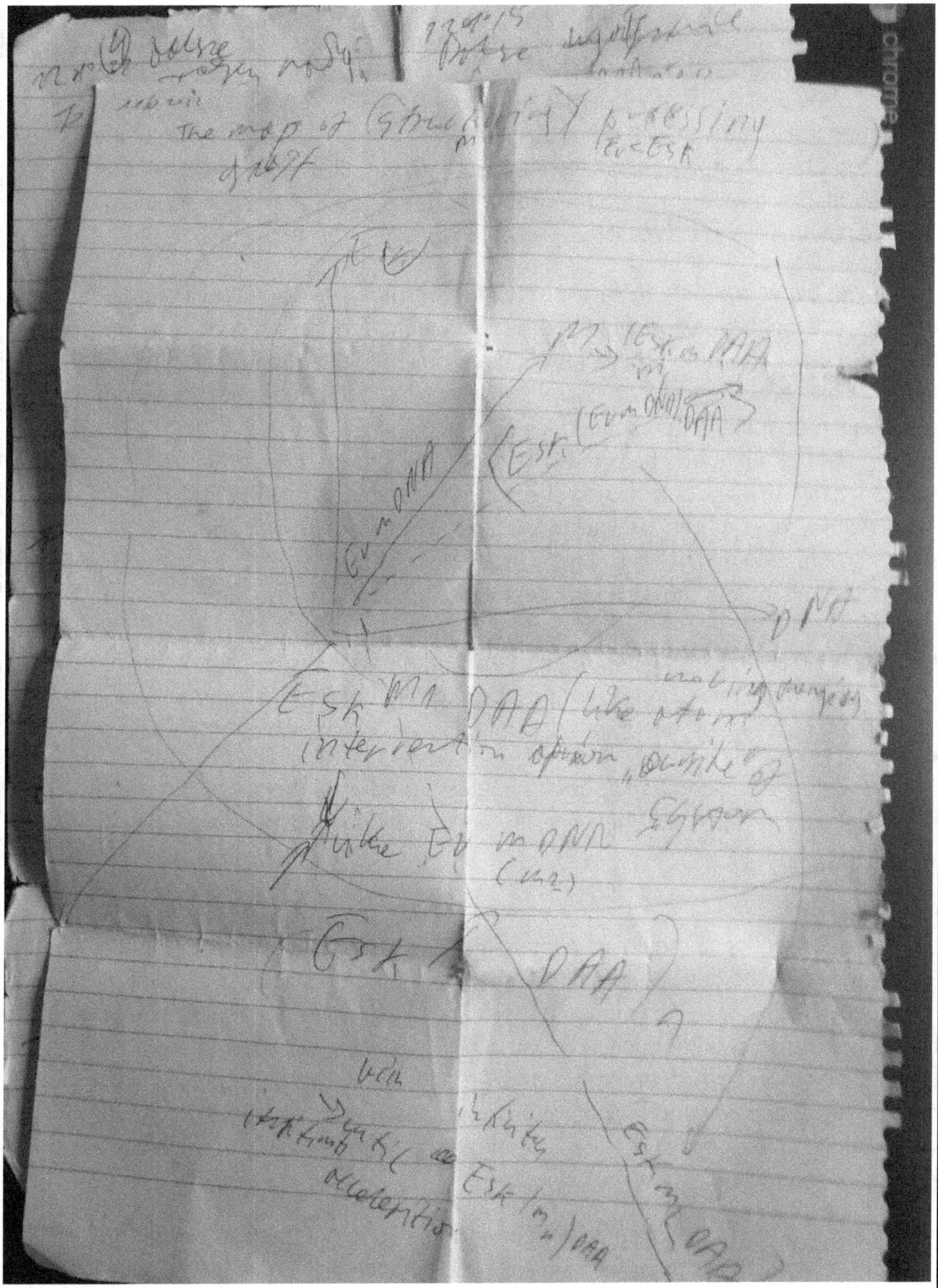

Cars with different m parameters of a given "material" level

Internal ones with the same external structures have different potentials, performances, the same with other objects, phenomena, where the time of the same supposed phenomenon has a different course due to other not always the same sources of structural components ... from different sides Eskm1daa+ Eskm2daa +(xtimes )EvmDNA +(x)EvmDNA...

Talk about hyper creation, replacement of the cosmos – re/evolution alone is not enough, even neoevolution, but hyperneoevolution, hyper creation of cosmos changes, as a democratic coefficient of contribution to change, as the basis of dynamic property, the trend of a given environment of the micro-macrocosm of matter -application of space macroeconomics
when creating, put matter before a fait accompli – and it is created /made automatically anyway, only how to direct it – as if from the outside, from nowhere, supernatural, over natural, replacing the existing perception of matter with processrooms Eskm1DAA = P = m2 < m3 = Eskm2DAA Pnm dynamic > mn static infinite in scale .... but mostly in its own fashion determining the direction, level, and space.
(...utility and desire factors) in pursuit of the matter of events, create our own process, matter, armor, material cage, event processroom through processing, initiating, stimulating processes without delving – because it is impossible to do so in-depth

Infrastructural, autonomous, process systems gigantic in relation to the existing element-> user-> next Neolithic revolutions -> at the cosmic micro-macro level ->

economies of scale
the macroeconomics of the cosmos intensifies the existing economic ... ecological levels – you need to develop the following tools of the neolithic cosmic, biological, and physical revolution ......... to win the modern economy to this cosmic economy building a horizon of infrastructure for yourself ....-> which will expand the potential plan opportunities DAA DNA -> property so-called expansion of the property of values through this expansion and thus in return strategies for the development of and so existence, i.e., life and matter – the face of matter -> processrooms process, i.e., open and not closed rooms

getting out of the ego's geocentric horizon ->.. freedom, perspective beyond the masses, existential and other barriers contained in incompetence, inefficiency, in... science quite unphilosophical, but physical organizational process action .... that infrastructure as an environment its properties and, moreover, the need for economy, efficiency with itself – resources -> so that the flow would be more and more in our direction, and at the same time not destroying the ground, but rather crushing it properly by processing

Increasing taxes means increasing competencies and working on the environment. ....increasing the tax potential of economic infrastructure, getting out of the classic market bubble -> increasing the DAA chain <-m1 <-EvmDNA Esk = DAA/Ev m DNAs Esk -> cost more on credit shifting indefinitely... costs-> burdening the environment bold on the other hand, but the infrastructure that this environment co-creates for... these taxes is connected with the cost burden with the cost ..environmental infrastructural loan

credit environmental burden DAA and at the same time efficiency potential Esk DAA-> infrastructure investments credits DAA DNA +DAA++ -> DAA expansion Tax at the expense of environmental infrastructure..... for further expansion of investment infrastructure.. tax hyper evolution in and out
Esk costs environmental efficiency .... credit its efficiency...
political, pro-fiscal, and credit, taking into account classical economics as well as physical, biological – this is the macroeconomics of physics, nature, environment, science, and medicine
do these Esk and DAA 1,2,3...
credit tax efficiency, economics, competence in the areas of further expansion as the basis for existence.. science/technology

application an attempt to apply this mechanism -> in the dynamics of changes
Esk....mn-> matter of mass phenomena objects projects ->DAA
more conscious, responsible, competitive ventures in the field of macroeconomics, economy, science, politics, society, space, micro-macrocosm

such an existential puzzle of the cosmos economic -> to be used, an attempt to replace this tool with finding this loan – more efficient
Esk,...investment tax DAA strategy is all in one -p> DAA and Esk
redirect – new spaces, liberate, redirect, but additional investments – allocate, liberate DAA->m<-Esk will release it – it will affect, energize, specify, give the basis for crediting the production of economic development ... in terms of the cell, atom, organism of the economic system, cosmic, planet, and other less or more open economic systems -> process

Esk->m1->DAA
the mechanism stimulating the initiative of stimulating the development of ecology by processes of nature with process rooms.

To be continued with going out from an airy position of weak laws to the new and old build/t environments parameters by bold initiative/algorithm.

**The greater the process, processroom share of Eskm1DAA in the infrastructural, non-structural, sub-structural background of a given phenomenon, the m1 = EvmDNA object, the more its so-called constant dependencies, properties will be lost in favor of new "ego/centers" of perception and expansion, including biological, physical ones. 1. c.**

Our task is to look at something other than the classical standards based on only horizontal, centric parameters, not so futuristic but more neo-creative.
Finding this always process/dynamic, non-material/non-ego/centric EskDAA for transforming reveals the subliminal properties of a given "m" of the environment's infrastructural processroom – in practice, the application of these macro-economic, macro-scientific, macro-technological relationships creating, perceiving, tearing out deeper, hidden coefficients affecting, to a given degree/level, external/internal constants relatively/respectively, physical and biological laws.

Credit/tax burden/for improving the environment – Esk ->m, investing in infrastructure – m <- DAA extracting, developing, extracting its more profound layers of potential for our development -> survival on deeper and further infra/structures of the microcosm.

more than the environment, infrastructure processes
The infrastructure/structure given to this environment may omit the so-called laws, properties of this environment, albeit not necessarily violating its existing parameters, the so-called material face, the so-called material envelope of the given environment, environment, existing infrastructure parameters, but ... changing the structure in deeper layers, infrastructure to structure and vice versa.
That is, the infrastructure and its properties are unchanged, but the internal structure is, yes, surpassing, disregarding the existing laws, diminishing them ... It's like the performance capabilities of a wood stove in relation to a coal stove, in the fight against cancer, precisely removing it, treating it without destroying other parameters. The same applies to other biological and physical processes, where they will be treated with harder/more complex inclusions – the classic parameters of time, space, etc., are omitted to infinity.
... It is a kind of procedural perpetuum mobile, increasing the place of our rights and existing competencies
Energy cascades beyond the cosmos' ocean horizon
Infinite process relativism
proposed on a shortcut through new
a microprocessor system as if on a narrower, deeper, more precise for forging in the current area, a kind of empty air, material, from which we let go with ... wooden hands, we cannot act effectively ... we are talking about taking over processes throughout the systems, changing the existing properties of matter, time.. ., cheating time, matter, actively not passively shaping its properties, accelerating evolution, revolution, neo-evolution, shaping back and forth systematically selected fragments ... process room .... being the basis, infinite in the plasticity of matter, energy..

Crediting, investing, developing crystallization/further processing, improving the material/existing processroom intended for environmental processing, contributing to the change of basic source properties, without necessarily transforming this material of the environment subjected to processing – something like moving a person faster, not necessarily rebuilding himself, but materially interfering with the environment materials ... car, rocket ..., heart valve, structural reinforcement of fragments of DNA components, not changing it but increasing its strength, the body's strength for its existence, possibly with appropriate external material interference, breaking the existing physical, biological, due to our depth intervention, as if changing the underlying

data of the medium in which the processed material exists, in time, in space. Often, very deep physical and biological interventions are needed to leave the material from the outside as small as a human, but to increase the chances of its existence by building a house, improving the home materially, in deeper and deeper areas of these macroeconomics and cosmic macroeconomics, economics, crediting, interventions – it merges propelling these scientific ventures, economy, supported by investments, credits, tearing from nature, but rather crystallizing tools more profound, further evolution, but with more our participation, knowledge, competence,

the inclusion of the economy will make it possible to economically but also scientifically support civilization, neo-evolutionary, creative processes of further processes of the matter and life, its increase in the spatial properties of human beings

Deeper and deeper interventions, observations of the structural and micro-macrocosm of the physics of biology indicate a systematic decrease in the classical parameters of space, time, energy in the progressing technology, the abundance of matter, the processing of matter, the processization of the process, until the current ego of classical, centric views of the cosmos become less and less useful.

There are no insurmountable barriers...

**Macroeconomic, macro scientific infrastructure/structural investments in further, deeper levels affecting neutral barriers, laws, trends that are supposed to reduce our potential for development = survival. 2. a.**

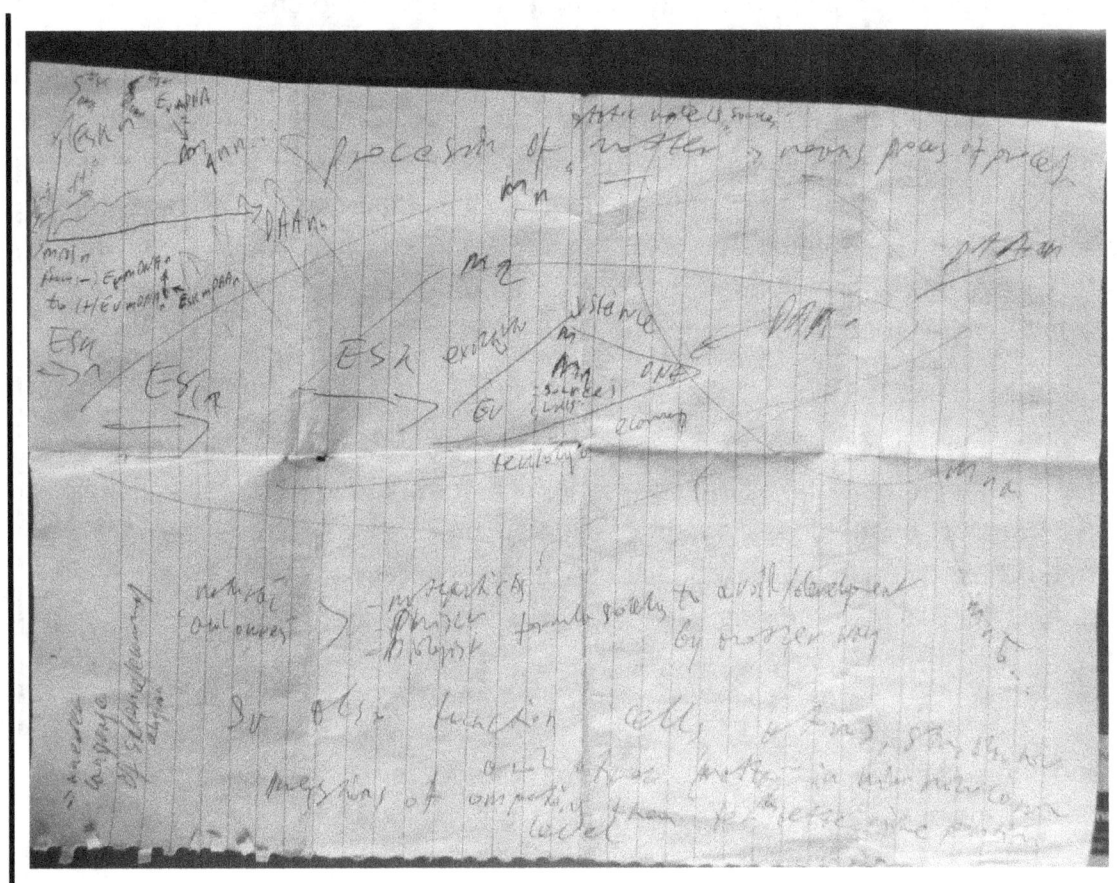

The matter is a process! This matter must be maintained and developed through this process to exist in it, even virtually. The process is a constant, constant provocation, stimulating changes, energizing/=effectiveness of this juggling of material changes, including and above all process, physical and biological processes at the instrumental level through infrastructural and evolutionary development; investments/credits, as the only chance of going beyond the existing states of biological, physical, economic, political and religious relations.

The changes are energy, efficiency, economics, crediting DAA with a macroeconomic and macro scientific investment tax in micro, macroeconomic, and scientific ventures...

Increasing taxes while increasing credit, i.e., simply increasing investment and constantly adding costs, but also infinite potential for development, pushed by lurking threats. Threats, regardless of whether we make these investments or not.. but this is life; this is how the systems of the evolution of matter and life function.

Without this effort/investment opportunity systematic, we have no chance; we cannot demand or point to nature and God.

Our process competencies, i.e., our contribution, determine our material strength in further breaking down barriers to development, which are the basis of existence in the classical economic, classical scientific field as well as this hyper-environmental field, which is to literally carve out, create, own new values in this environment which by changing have the current sense or lack of sense of parameters, classically counted matter, time, speed, sub-properties of DNA, object-physical, biological values.

These economic terms of the phenomena, i.e., responsible, creative, divine, evolutionary, are more than the scientific, economical formulation of the topic, which has been taken over by the development of maintaining one's position on at least a neutral, middle level, in a dynamic, equally economic environment from the inside and outside, is the basis for thinking, AI language..., action, planning in classical and non-classical

manifestations of activity, man, ... stone, star, atom, and other smaller or larger, known and unknown participants of this race, this infinite flight/explosion of the microcosm.

Redirection, i.e., participation in directing the evolution and development of systems, often we do not know them, with invisible players of our known and unknown environment.

• such blind advance secures, tracks, infrastructure inside ... and outside, and other participants more or less alive, active, – built into its call, inside, into structures with us and interacting in the cellular, atomic and other terms larger or smaller economic organisms -> process room, in the process room of the micro macrocosm

i.e., further investing in the infra/structural infinity flow stimulation mechanism in the macro and microcosmic direction— rather microcosmic as a tool of macrocosmic engineering
Esk->m1->DAA->mn

a subliminal, supermaterial process... because matter by definition has some barriers, it is a cage, a mask on the basis of which it is impossible to draw final conclusions – possibly on some symptomatic section.. yes
but process competence investments increase the potential of properties indefinitely in each case – they are the foundation, the determinant of the properties of matter, which is the micromolecular image of any process -> process

Esk credit and DAA taxes decisions infinitely determine the m1 process – the foundations of matter of a given level of interest to us – this determines retrospectively
m1=EVmDNA

Sense Esk credits, DAA taxes investments plan process strategy infinite because of infinite lawsuit potential, hyper neolithic revolutions.....

this is not the ultimate transcendence of God, nature – it is the support of the one we know; against threats and at the same time against centric ... materialization ... any barriers to development ... fatalistic relations
it further broadens the horizon in every field of activity -> processing

will pull you out of this kind of gravitational pull of the material
-> this is not some surrendering to the course of nature, but quite the opposite
it is simply an indication of responsibility, competencies that are the basis of matter .... in the development of the existence of each of us, the environment and ... God...

# Economy, i.e., awareness in non-passive responsibility beyond the ocean horizon/spectrum (DNA) of each process structure, sometimes

# expressed in images, experiences of the so-called matter of phenomena, as the basis of any functioning, life. 2. b.

We're moving on. to specify the meaning of these digressions, which is the co-participation in the economy of process relations for the constant preparation of the ground for even more complex science, more challenging economy – macro economy – macro science, for the sake of completeness, frontally the whole background, the basis of the micro-macro universe – not from in particular ..., some random, but a detailed and massive systematic takeover of DNA bases as a process tool for controlling matter and life in the environment, on the investment path of increasing potential and opportunities, artistic and other options on an increasingly civilizational, over primitive material, fatalistic background, a spectrum of events and phenomena.

The proposed assumptions are also helpful in every classical economic, scientific, artistic, individual acts of a man, nation, and other systems and civilizations .... in revealing, shaping a universal scheme overlapping with classic biological, physical models, intended to be more opening mentally but physically closed.

That is, beyond the geocentric system of thought, the sense of modern civilization as an investment way of processing, which is the basis for any further changes by increasing participation in those projects that are the involuntary basis of development, i.e., life,

are the only alternative, chance, potential, in redirecting all flows known and unknown on infinite levels, which also have the so-called subliminal share in shaping the properties of the so-called material, the so-called laws, closed in some process-material niches, where we are more or less hooked, ... we (our known world) like air, the cloud through which it penetrates, affects more or less. Our task is not to be so...empty, but as it was in history, to consciously increase our share of responsibility, not by chance. Will this cloud disappear, or will we, more than empty stone or steam, have something to say about deeper and deeper further sources, forces, affecting the environment and us who are part of it, It's not a passive, but a co-perceived, co-sharing attitude and behind it infrastructure. ... then we can say that a tree, a stone, a star, a galaxy, a quark, a photon become our tools, building blocks for ... e.g. hanging on the ear as an earring.. not harming ourselves and them... as it used to be sometimes, it happens in our history, in the present – let's be more aware of these economic factors in which we are shares, let's make this earring smaller, let's have less furniture, a nice dress, let's have a galaxy as ear piercings or a hat, let's have a healthier, longer life, longer life of the sun, .. the shape of planets, galaxies, i.e. asphalt and milky ways... let's have this micromacro perspective of the cosmos just more comprehensive, in a more conscious way processing plant, investing, infinite growth potential which we are forced to use, to be already pushed through our stagnation with the endless potential of degradation,...

Pocessing infrastructure as a not passive driver of any property evolution in an infinite spectrum of the potential of processing as the foundation of any matter...
(economic)processing spectrum of matter
As new over ego-, geocentric passive or active (depends on level) participation view of the universe of matter and life

• economy of a spectrum of processes
of actions/process rooms/"matter"
driven by ever deeper levels
It's about exploring further, deeper integrated circuits.
The spectrum of seeing, acting, processing determines the laws closed in a given range, but also beyond, in terms of the so-called phenomena of external forces, internal, but also our process, strategic, model, not so geo/egocentric interventions, if we want to go beyond the laws of the so-called quadrature of impotence.

Hyperspace investment subliminal factors of
Emc2, cancer, life expectancy, etc.
Over fertilizing treatment of one's place in the cosmos Beyond time beyond previous
paradigms...

This Eskm1DAAn tax credits system is the most effective,
The most responsible point of view in the field of infrastructural interactions with the
environment at the level of expansion, development, and, thus, climate protection.

Investments in deeper infrastructural layers of the environment, beyond the classical
economic barriers
Applying the investment credit tax in the cellular, nuclear, and global rules of their
functioning, development, and existence. The indentation of this economic or ecological

mechanism into neo-structuralization at subthreshold levels that do not directly affect
the parameters of matter and life at a given micro and macro level

Provoking changes energizing (increasingly better and more efficient structure
accelerates back energy efficiency for further massive transformations and so back)
changes through
the process is somewhat less coherent than matter.. but the matter has some briers,
less ecstatic images

The process competence of the investment will increase the properties. They are the
foundation, the determinant of the properties of matter (as a further crystallized, frozen
version at a given level of processivity; further on, it will always occur under the shell of
this material experience/image

, matter, which is a microparticle image of any process
ESK and DAA in terms of taxes (more outstanding and more effective share of capital
and infrastructural macroeconomics, which results in lower and lower credit costs,
respectively, more effective investment decisions, and perpetual motion growth) ever
smaller credit loan determines

ESK loans and DAA investment taxes plan process strategy ->m.
Do minor lawsuit credit because of tremendous lawsuit potential
The impact of neo of the neolithic revolution changes this view of the economy,
physical, political, medical, and religious, which is not ultimately against god nature – it
is supporting at least the one we know against macro-microcosmic threats and material-
centric ones. ... a primitive materialization of views about any barriers to development or
fatalistic relations.
This further shifts the horizon in every field of activity – processing
an extract from this specific material gravitational pull... but with process bases.

It is not some theory but simply an indication of responsibility, competencies that are the basis of hope and development – that is, the existence of us, our life, and its environment-infrastructure, on which it can develop the guarantee and quality of its being.

Don't look back – petrified processes – a reality that doesn't exist anymore, but to fight the progressives away, these material fossils – have an impact but only in m for EvmDNA=m1 for Eskm1DAA.

There is no choice but exploitation, explosion but possibly controlled – it's a matter of this extreme boundary between stagnation – not copying, and development as a chance for better infrastructural foundations of development = existence -evolutionary creation but the not pure rebellion of systematic hyperproduction.

Confluence of an investment's Esk/DAA investment/credit function, sometimes more investing, sometimes more lending
application, finding these health benefits and investing inspiration. I call a credit facility a driving force because investments create the following foundations for the potential of

sources ... and vice versa
DNA lender creator (by cost) of ecological change investments
DNADAA creates/credits neo-ecological/evolutionary changes subliminally

The ecology of managing the flow of a team of processes, processrooms made visible materially ... but always at the highest level of processability, capacity, building/crediting potential – outlined, unfortunately, often already dogmatized "ego" materially in the long run (as it was with geocentrism, and it is with other dogmas "ultimately" determining the sequence of factors of physical, biological...

# The hyper economy of neomatter or nomatter/"matter"=processrooms... bey ond the horizon of "natural" evolution. 2. c.

Macroeconomics of the environment, i.e., systematics science of the economy of taking over, extracting, processing, neo-evolving, i.e., more than revolutionizing, simply producing, creating new infrastructural foundations giving new foundations generation, i.e., new DNA, i.e., more dynamic than DNA administrative activity of the Eskm1DAA process integrated circuit as the basis of all phenomena, tendencies, activities, and properties – under processing expressed later in structures, tools for further materialization, neomaterialization, the cosmos, taking over, replacing, mastering its

structures/facilities in geometric progression – because only such counts with equally fast other progress, forces, sources of the micro-macrocosm.

This economy needs infinite system support, understanding in this hyper cosmic creation = investing, where this unlimited process potential is to be used in the process resource management mechanism, which credit is their expression, adequately dosed, directed, even .... credit(recognition) plus interest,
You get, for example, an additional, e.g., + 1 percent extra relief for zero credit. For instance, for a house for 1 million, you pay back 990,000; the same applies to any projects of phenomena at the microcosmic level – we drive our own trend, hyperevolutionary boom ... with hyper economic, hyper ecological investments, using extra pressure evolutionary, self-perpetuating, .... macroeconomic investments multiply multiplier microeconomic, and so in turn ... a kind of ecological-evolutionary pushing ... race of evolutionary micro-macrocosmic armaments.

This economic democracy of infinite potential for the procession, crediting trends of a total transformation of a global, cosmic, hyper-evolutionary, neo-ecological character, where responsibility is expressed by active, not passive, participation based on responsibility in contributing to one's immediate further benefit as well as the environment, which is a source of strength, material for further processing into ever-lower/precise material, structure democracy of hype-revolution

involuntary responsibility as a participant, an element of this environment, a system of forces of mutual processes, i.e., energetic transformations, which translate into the conversions of the so-called matter into low infinity, according to the directions of those who give a more significant contribution to this process and the energy of transformations as the basis of matter and life transformations on a scale that even the greatest fantasists and futurologists unimaginable, because it exceeds all limits in the control of all phenomena, objects, the cosmos, and the macrocosm

democracy as not a populist but a real competitive participation in managing your shares, requesting the right tool for this transformation EVmDNA= m1=ESK/DAA -> Eskm1DAA= m2 =...
we are involuntarily responsible for these patterns as if we were internal parts of these physical, biological systems because we are them; the issue of proper referral, constant determination of process competencies in this as a shareholder of input holder of matter that is ineteresting til us

Democracy of environmental competencies DAA – if the environment and other systems are more mobile, their contribution will translate into perspectives, the spectrum of this environment

It's not about gaining survival but about replacing the whole "m" of the macro cosmos' environment
replacement is not revolutionary but neo-evolutionary
for more and more complete changes in the properties, structure of the infrastructure – we are the process infrastructure of matter/life; as long as we are active in this topic, it will be more flexible for our needs and the environment

Esk – EvmDNA -DAA Esk/Ev – m1- DAA/DNA = m2 etc with m3, m4, mn

the resulting – from our DAA or other DAA systems – infinite levels of the size of the processing structures, which include and do not include the contemporary technological parameters (spectrums) of perception and the operation of counting time, speed, volume, mass, structures of process dependencies ... structural relationships of the cosmos is to be used

the question of the pace, direction of self-propelling control of this process
when concerning the environment, we can put up resistance to increasing activity, processivity, and the precision of "permanention".

The question of the speed of hardness impacts of our construction of process structures, process rooms

Because we are like air, almost empty space in relation to the undiscovered structures of forces, deeper and deeper building materials, blocks, with which, to a finite extent, the layers of the micro-macrocosm are dynamically composed in a process-like cosmos of microprocessor bricks, microprocessor, integrated systems, each of which is an essential element for other m including our EskmDAA

Material process evolution from wood to atom to... material processing
future processing changes of our partition, participation, so our real/absolute/actual position. Neo algorithm neo DNA – DAA...not on material static, but process constant variables...

To continue with the hyper economy and neo matter development by... + taxes and – credits investments for mega science, mega economics, and mega life by/for mega infrastructure/environment...

# The mechanism of DAA competence Esk grows by the P neo-processing of m cosmos/matter. 3. a.

Economic democracy competencies, environmental competence -> DAA tools.
It's not about conquering or surviving but about something much more. It's about replacing – not revolutionary but much more neo-evolutionary – all the encountered m – cosmic environment micro-macro -> further changes in systems, structures, infrastructure of the matter of life – for this you need mega tools, mega organization, steering, decision-making room supported by robots, AIrobots, specific levers but which

can be dangerous, or dangerously, each process can backfire, reflects the issue of sound economics of this process, so that the benefits and development exceed the risk of degradation. It is a constant dynamic process, which classically cannot be fixed materially. Only economic and economic directives define the meaning of the activities of deeper further systems, which also constantly develop their DAN / DAA competencies. These foundations generate changes in the internal and external environment for a given system/scenario.

This mechanism has already happened, it must be conned into cunning, but at the same time, strategic, responsible actions – idleness, indifference, ignorance of this economics of the systems of the matter of life determines the chance for the potential of existence, development of civilization for each individual – because for this you need competences, foundational improvement, not theoretically. Justifying some existing system- there is no anchor because it was created in our non-competitive, passive neo-evolutionary attitudes...

Not on material static but process constants of process variables of development and force in relation to the dynamic environment of existing properties within us and beyond

Perspective, a strategy of action, processing -> better organizationally precise reaching further sources of processing, creating own seeds, sources of metabolism reevaluate the properties, energy, time, biological, physical performance at an ever more profound level of the micro-macrocosm.

The legal "reality" of physics, physical and biological perceptions applies to a given level of "imaged matter" processes.
There is no sure guarantee of an absolute law unless limited to a given level of the processes of development of matter

..., not an absolute reality, because it is changeable and which is infinitely hidden in an infinite network of process systems -> this law is the overriding hyper-relativistic, supra-realtivistic law – we can mark it even more than relativistic laws/directions... The law indicates the infinity of the potential hidden in the processes and... processes we initiated ourselves. We can continue to stir more systematically, economically, productively, and responsibly.

Energy... is the force needed to survive, to process the matter necessary for the reprocessing, neoprocessing of further matter structures, and vice versa/reverse definition of mass, light in the so-called "air"
our world can be like air where other forces, even with the same so-called laws, can be indifferent as well as competence can transform our world, despite creating our world with the same laws/rules, but on different dimensions of the process of materialization. It's like in the days of carts when people talked about pollution in New York with their excessive use

cart strategy, rocket car from... openness to space total material development strategy speed up its space economics

Creating a situation in the so-called hard/complex environment of systems – systems will be softened for ... air by harder, directional, sharp, more precise "smaller" tools for

this process intervention, process room systems
breaking through the barriers above the laws that are true for soft technology and economy, ecology, the evolution of processing ... but insufficient, sometimes contrary/contradictory for harder/more complex technologies, eco-economics, which will be harder/complicated than perception, tech laws of the current environment

the need for changes going beyond the event horizon rights -> manufacturing, directing the development of technologies economics ecology evolution ... energy further processing improving processors
-> the issue of creating a more efficient organization, algorithm, but also structures of organizational and economic competencies, so simply changing the infrastructure of the

environment of the structure that we want to break through, be tougher literally in interfering with structures/facilities such as stone, iron, atom, ......, more solid process or ... materially new space new materials create a new/unique quality, properties in biological, physical, economic, evolutionary and ecological spaces.....

the so-called new explosion and reversal -> creating new matter, new quality -> their levelness and depth
the depth of intensity of the processing plant gives experience as a paradigm above all laws...

To be continued in investing, neo-integration – supra-system management systems (A.I), managing the process bases of environmental infrastructure.

# Accuracy determines all subsequent effects of the mutual laws of objects subject to classical physical and other legal relations with each other. 3. b.

Mutual competition expressed in being tougher in sculpting our space. Hardness, process directionality, i.e., mutual, increasingly resisting barriers in determining the course, infrastructurality of structurality, and then further infrastructurality, which will be the basis of process structurality, and so on, and then materiality on one spatial (cosmic) level.

Accuracy determines all subsequent effects of the mutual laws of objects subject to classical physical and other legal relations with each other.

This process infrastructure is an investment in the process basis of material structurality.

These investments are a tax credit that we work out to raise the level of generating structures that interest us based on this process of determining further "material" events.

It is the basis of life = the creation of matter or life.

These taxes are always for the so-called public or infrastructural investments for a set of structures, including life process structures, which is the opposite of saying that taxes and death are the only certainties.

And there is a lot of truth in this, only often misunderstood fatalistically and degeneratively, and here it is about a completely opposite process...

Without taxes, there would be no death, and it is rather the opposite ... thanks to taxes, we can postpone this death indefinitely ... that is without infrastructural investments, we are not able to shift the environmental burdens that occur with the symptoms of life.

Life is constantly moving in the environment's infrastructure to make it function better. So there is a need to increase interference in the internal and external environment of this life of this process, which is just involved in this general environmental processes with infinite potentials and threats, where the infrastructure tax investment on it has these aspects of environmental platforms of life and matter processes to work out ... that is, a tax to live, that is, without tax, there is death, that is, there is no tax without life, there is no life without tax. Tax/developing infrastructure of a structure and death/destruction are contradictory tendencies.

Infrastructural systematics, specific clarity of process crystallization directs and so materializes phenomenal effects more than randomness.

It gives more significant potential, chances, infrastructural basis for determining processes, phenomena that interest us if they properly reconstruct them in their own direction.

The more systematic the crystallization and the structure of the process, the greater the force of penetration, literally and figuratively, into the process structures of the tendencies of matter and life at the Astro/physical, chemical, biological, medical, economic, macroeconomic level, etc.

This kind of systematic displacement of a given space of the structure of processes, intrusions into ever deeper selected segments of the micro-macro universe of physics, biology... creates -> supports

, sculpture energizes, builds the birth of another world, simply a new child, a new generation in our activities and processes.

Simply taking an active part, that is, merely active life, the growth of life – because without this giving, there is no life, no basis for life, no rights to life, competence to live, and other so-called physical and other laws.

A new DNA competence base constantly developed by action -> DAA is no longer revolutionary but much more responsible, i.e., neo-evolutionary process matter as the basis of life and any action, action, counter-reaction action.

As a constant redemption by evolutionary economic, creative, ecological. The rights of competence to create new, more efficient foundations Ev -> Esk.

And just if the existing skaldic blocks of the structure are not sufficient in the requirements of the parties of the micro-macrocosm environment co-responsible with the appropriate forces, then this new tearing, carving is to give new foundations

competence in the field of ... the life of matter, cosmic medicine, etc.

One should move the development fronts further as a guarantee of security, existence in a dynamic environment of mutually pushing micro-macrocosm of systems.

Simply responsibility and competence for what is done and not done is the answer to the basics of anything, ever.

There would be no overheating, no cooling down of the economy of this hyperevolutionary and thus also classic because automatic infinity ... infrastructural dynamic supporting development goals in infinite strength and pace amortized, supported by autonomous AI systems for redirecting, directing the processing energy possibly suppressing in some cases.

Investments are processing in processroom frame changes..-> credit is the opposite of

consumption -> they are not aimed at comprehensively strengthening the foundations of existence, development, but

credit investments, taxes for further subliminal strengthening ... but also material, economic, matter, and life comfort of the position...

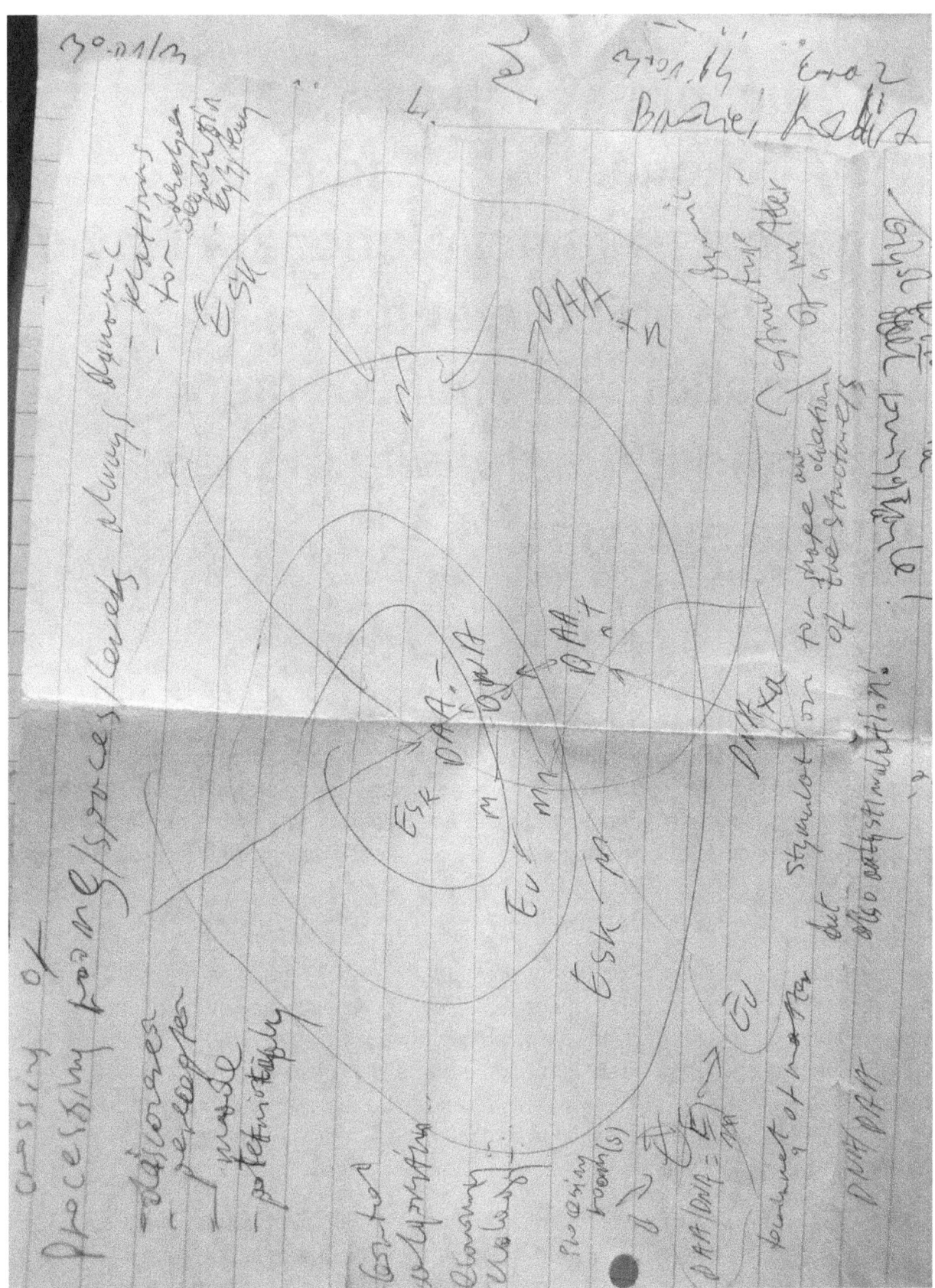

# An epic, pre-genetic as (r)evolutionary way and hyper genetic as a neoevolutionary way of accuracy of intervention, cooperation, co- management throughout the density of matter and life systems. 4. a.

Let's start with further outlining, gradually specifying ( ... but not so much giving a margin of initiative to all interested parties) the framework for developing an algorithm of the mechanism of purposefulness and efficiency of infrastructural projects increasing the potential of a given structure caused by its transformation with the help of external infrastructural projects that need a self-sustaining growth mechanism by .... reducing the costs of crediting these "microeconomic" classic market projects, and at the same

time increasing taxes to support the infrastructural growth of these perennials – external intervention at the classical (macro)economic level of the activities of the economy, physical, biological, astrophysical, medical, cosmonautical environment , ecological with a higher degree of intensive transformation of the entire micro- macrocosm with possibly more minor side effects – smaller than the benefits of transformations on a (macro)environmental/cosmic scale, macroscientific – going beyond the existing barriers, development visions.

That is, it is about a new deal of the infrastructural development front above the market, classical economic, as well as beyond the physical, biological, astrophysical, cosmonautical, medical, intended to control, rebuild entire microstructures of the cosmos of matter and life in geometric progression – it requires an increasingly efficient used artificial intelligence, robots, chatbots...

That is, in the neo/evolution hyper generation investment
it's not about small observations or epic preventive interventions, but about dynamic hypergenic, neo-evolutionary, neo-infrastructure systems of reconstruction, taking over by encountered structures, building infrastructure around them – i.e.
neo environmental, neo infrastructural economy of economy, astrophysics, biology,...medicine, aeronautics, etc.

The proposed method of pushing, launching the entire economy, physics, biology, civilization beyond the orbit of attracting a tired system of actions, thoughts, strategies ...or rather their lack, by assumption, by attitude.

An epic-, pre-genetic as evolutionary manipulation and hyper genetic as neoevolutionary manipulation ->As hyper evolution of matter, life, and vision.

The steady increase in cheap credit (which indirectly contributes to the growth of the infrastructural platform to further drive the classical economy, and then science and vice versa) of investing pushes, shoots
capital, the platform of the matter of life involuntarily as a niche, where the existing patterns of action, classical and other processes are translated into a given direction of change, a specific lever of processing, investment, credit (in mutual self-propelling macro-microeconomic ventures), adequately directed, increasing the direct and indirect bases of matter/environment/life and other more or less benevolent physical, chemical, economic ... economic projects in this direction as an interfering basis, intellectual basis, but also another more or less abstract basis, the so-called artistic, not always coming out of the physical and economic calculations of a given level of proposed interventions, the development ones of these evolutions) can be so-called unpredictable (human freedom of choice and a given set of forces – not always ultimately obvious) – because decisions, factors can result from deeper unknown always! potentially infinite decks at first glance and the so-called classical economic, scientific coherent ... but in reality, never inconsistent in a longer or closer perspective/spectrum.

Because of this infinite potential, infinite energy is in each of us, in every element of matter and life, in classical and alternative terms. It is the fundamental law of relativity,

or rather ruthlessness in vectorization, participation in the emulation of processes, cunning shaping, creating matter, life, qualitatively, quantitatively...
The goal is not self-fulfillment, but unconditional conquering, more or less economically, responsibly, artistically of the cosmos – through its more or less interventionist, economic, artistic transformation through the infra/structural ladder. Not being enchanted by one step of a given ladder, barrier, platform of potential = degradation, but

the infinite potential of diversity .... we are crippled, but let's give ourselves a chance ... get out of this disability geocentrism of the infrastructural micro-macrocosm of matter, i.e., increasing the life potential of its base ...

Of course, we are talking about nothing extraordinary but about the natural, ...obvious driving of the structure of infrastructure
and so on interchangeability in evolutionary and hyperevolutionary thinking and action. It is about the efficiency of the awareness of this process, which has an economic, religious, divine, natural character, because every (un)intentional opposition, ignorance is a suicidal impermeable nihilism.

Each next infrastructure, tool, process, and its model is a supermodel of the previous ones and a submodel of the next ones. Still, it can influence, as if from the side, invitro, from beyond, the matter of the process of the object or issue we are interested in. It is a tool that raises above the limitations of a kind of illusion of the existing world of infrastructure->structure, a tool and thus the idea of strategy, and a more or less ignorant or more active approach to the surrounding factors, which are not some independent set of fatalistic so-called "constant" natural systems.

To be continued on the tools of change in mind/action-> process, i.e., hyper DNA/DAA or Esk/DAA=m...could be found in any physical and other formulas or models.

# Magic of matter->process or from nothing to thing...processing makes any matter. 4. b.

Further attempts to determine the processes, the mechanism of matter creation, or rather the structures of matter, or rather the structures of the processing plant – the process plant mechanism, i.e., a kind of juggling of matter, or rather a set of processes, i.e., creation from nothing, a sort of micro explosion ... infinite micro Bigbangs system of creation-neoevolution – this is the economy this production plant of the cosmos, matter,

... life, i.e. process responsibility for the image, shape of matter, phenomena that are the result of all this.

Everything where processing is a material or rather a process, a civilizational evolutionary springboard of humankind in general known and unknown cosmos not hidden with a show, some game, but a successive takeover of infrastructure, creation of its own for a given new structure, which will be the following infrastructural process mental basis tool for a further explosion, exploitation, transformation of the existing values to an infinite degree.
And this is a response full of faith in its non-passive place in the cosmos.
It is this active ecology of conquest without losing yourself unnecessarily in fatalistic passive thinking, ignorant strategies. It is the pattern, this genuine faith, this fundamental truth about the dynamic picture of the world, in which we participate without obligation whether we like it or not.
This awareness, fuller awareness, responsibility, true evolutionary faith gives the proper shape to the divinity of our actions because this is the divinity where the god is hidden.

It is not a philosophy but simply production, economy, management, responsibility – I emphasize – full responsibility, where the boundaries are defined by our ... these beliefs and not illusions, images of what we have only seen and experienced so far ... these proposed patterns go beyond the limits of this scientific image, they are this religion of progress, they are over the scientific paradigm of the strategy of existence, development of us and the entire known and unknown environment.

They are anticipatory, predictive, as a friend of ours once predicted the existence of black holes.

P the process with Ev/DNA =m (EvmDNA = m1) is degradative, and the process from..m1 – Eskm1DAA is progressive – that is, the process is not equal to the process from the point of view of our developmental/=existential interests.
They p= EvmDAA these progressive processes determine the proper hyperevolutionary infrastructure projects for the following levels of structural potentials...

Processing can be a roundabout way
produce any materials and phenomena through infrastructural-> <-structural overlapping ways/forms of development, the evolution of ... matter, life phenomena ... e.g., from gravity to ... light, from carbon to energy, from energy ... to carbon. As entirely new artificial structures of the facility and, in turn, processes through this infrastructural structural outlays, in acceleration depending on the share of the existing directions for these so-called infrastructural investments(credits/interests rates down, tax up ->...), further increasing the further potential of increasing the outlays.. unconditional economic growth at the classical macroeconomic and physical macroeconomic level (macrophysical), biological (macrobiological), (neo)evolutionary. These transformations can be progressive and... regressive (non-degrading) but restore a given platform for physical, biological relations, e.g., life, fighting cancer, etc.
That is, a kind of pure manipulation of the course of processes, i.e., in simple terms, also time and other physical, biological factors ... processes
Appropriately controlling, algorithmizing accelerating the increasing potential of these

processes of matter into infinity, exceeding all observable parameters
them from the overlapping systems of their structures, infrastructures, which in a roundabout way from not the same visible infrastructural, structural position, but from elsewhere crossing but calling to these circles of structures that we deal with, direct ...

From there, these prayers and miracles are the results of these overlapping circles of unknown deeper areas of the procession.

This process basis of matter can be called a miracle – enchantment, creationism, a big bang every time because it comes out of ... nothing, i.e., a process but based on other convergent ones shaping the image, specific layers of this action, i.e., material, but not static, because based on an infinite set of past processes and future of their leveraged potential on existing facilities, effects, contributions of their DAA and other DNA based business activities Esk,....invest tax credits for these bursts...

Processing m1 is the so-called possibly non-cost
Investments, the system of periodically influencing changes in ... m, because this is the logic of economic evolution, the sense of explosive creation from nowhere. From nowhere, the next stages of development, i.e., existence, because in the process stagnation ... everything disappears, degenerates, dies in the p = m system, but not in the p -> m1 system ...

# Without infrastructural taxes, there is no (matter of) life. 4. c.

If we want to change the micro-macro environment (it is constantly evolving, it's a matter of our contribution), then we have to change the micro-macro infrastructural, which is the basis, any source of further changes in the structure of us – the entire cosmos in us and beyond, which of this environment infrastructural=neo-environmental superstructure -> supra-environmental process – because the process, initiative is the driving force of any existence, functioning, detachment, maintaining one or another

status quo, even the current, historical or future one. Ignorance, expectation, entitlement, and lack of development of foundations is submission to control other exploiting forces, which will not necessarily set the trends of changes or their "deficiencies" in our way.

The more often attempts are made to make these changes, transformations, the accelerator of pushing, pushing changes, pushing changes, building (neo-natural) infrastructural, the greater the chances of imposing pushing, redirecting current trends, phenomena, objects that interest us in a purely business, purely scientific sense, but more often it is about for a purely expansive approach changing global, cosmic trends on an open scale, towards micro macro totally and infinitely.

The process of twisting each reality, ... properties of ... "matter".
Material without any limits. The ratio of real (power of tax, interest rates p-> reducing idleness = death) process investments Esk to consumption investments Ev (in the direction of over-economic, over-scientific), infrastructural superstructure, which will become a new component to the structure of the existing environment.
There are no so-called material barriers here because this processing has an infinite character...
a constant procession race of first come, first served
the strategy of processing to the extent it is supposed to go beyond the current neutral processing, which was not, is not intended to change reality, i.e., consumption, opposite

to the hyper investment, neoevolutionary, hyper economic, hyper globally, hyper cosmic initiatives proposed above because directly aimed at changing the economy, economic thinking, i.e., responsibility on a scale ... an infinite scale, because it is, or rather, there is no scale of challenges in the classical or neo-evolutionary economic, scientific, physical, biological, ... artistic, religious terms.

The way to this is to build a production and economical tool infrastructure ... from crushed stone etc., which is to be the base of the following levels layer of infrastructural eggshell shape/think(DNA/DAA) of structure
i.e., infrastructural pushing of undesirable processes and stuffing of desired ones. These pushing infrastructure investments of a process nature are to transform the environment, which is just a component of processes fully subject to all laws, what conditions we will find for them and what we will also introduce, according to this neo-infrastructural construction
investments, taxes, credits/= Esk DAA, which are the basis of the strategy, development, or degradation concept concerning the dynamic environment pushing m->m1,
for an environment in which even the classic parameters of DNA, material, even primitive consumption passive behaviors enterprises and processes Ev, will be better than the previous ones thanks to the infrastructural superstructure, by increasing the existential, artistic rights – ... loans with low-interest rates = high support, which at the same time increase the potential of this artistry supported involuntarily more excellent bases of tax revenues to secure it (this loan!) of this current model and at the same time effectively push our civilization, our life further, because this is its basis, everything else is the philosophy of strategy of the fatalistic, primitive look of artistic vegetation, the self-destruction of itself and this environment nature that we know even on a far cosmic level.

These credits, taxes, and investments are not only literally economic but also in the light of initiatives and processes in which these economic functions occur simultaneously and reciprocally buzzing a given system.

Because without infrastructural taxes, there is no matter of life.

In the microeconomic micro science approach, it can be written that develop, take possibly effective loans (from nature ... from yourself infinite processing potential as an unlimited potential base of matter resources) for your infrastructure/structure, to further develop on an increasingly higher level of security, as and simply development, and then existentially, as a guarantee of one's living environment, it's further crediting/guaranteeing for covering infrastructural investments at the micro-macroeconomic level.
It is the right strategy for classically business and classically scientific development...

and then an artistic, civilizational, ecological, evolutionary, cosmic, guarantee of life on a global and personal scale, revolutionary, or rather neo-evolutionary perspective of each individual, us and the cosmos literally – I emphasize literally and not philosophically – this is an attitude of responsibility, which is the basis of the true faith and a chance to achieve God's goals, our goals, the goals of responsibility literally for the entire physical and biological cosmos, as a basis for building permanent protection of the development of life and not rejecting it on – not contempt for life in reality, which is expressed in deeds, surviving the sin of omission, passivity, ignorance, shifting the blame and tasks onto some external/internal forces literally on the shoulders of future generations.. credit/investment / this tax for life from life without coverage

• and we need to provide credit, guarantee the infrastructural foundations of existence and development today…

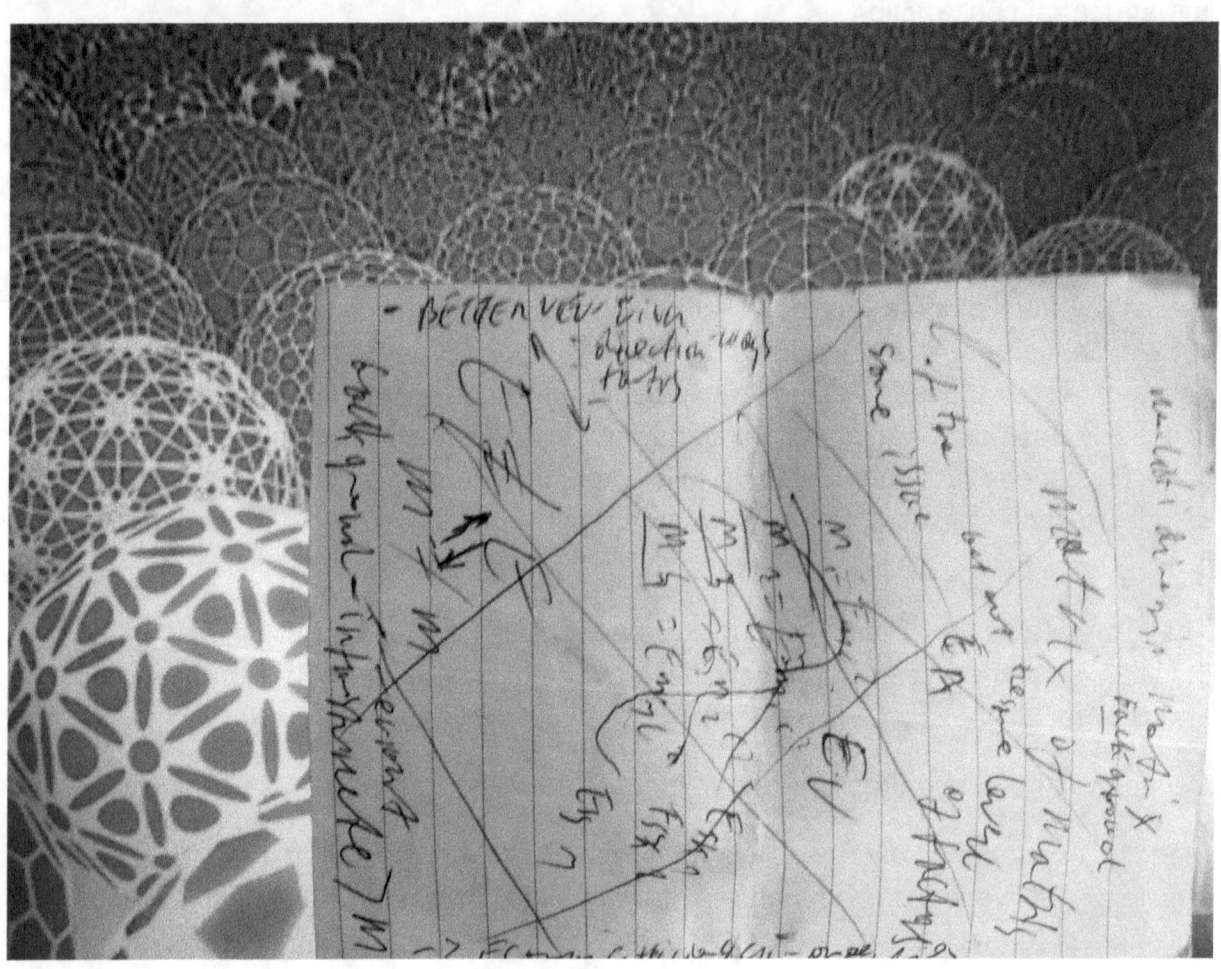

# Processing pyramids of matter. 5. a.

To what extent will investments->taxes, credits for this release in the infrastructure of areas above the current level of interference in nature, above natural, neo-natural, neo- evolutionary neo-environmental, neo-economic or neo-macroeconomic going beyond the classic market (artistic creation -> processing without regard to environmental conditions) behavior, interest aim,

literal and absolute systematics, algorithms of imposing transformation procedures as a substitute for ever deeper, further fields of matter (in this case, not matter anymore, but observable, experimental symptoms of environmental processing, its reactive properties); this is how the vital potential will increase today and in the future -> m1. We are talking about m1 and not m because m of the matter is inherently degressive in nature, and m1(n) is inherently progressive over expansive in most cases stimulating – imposing process relations to achieve goals ... a roundabout way beyond the current, or rather material, static process approach, beyond these static arrangements of the hitherto taken shots in favor of the constant dynamic, literally ruthless

chase, race of processes (as if micro big bangs from nowhere) producing bypass environments that can more precisely
(here is the key to co-decision about the processes concerning our physical and biological environment on the macro and micro scale, or rather the lack of scale in this dynamic proposed case) in the structural space to an absolute extent to the current system.

The DAA of processivity of the leverage of processing ->matter-> as the basis for shaping the control of neo/evolution gives a lever fulcrum (Give me a fulcrum, and I will move the Earth! Aristotle).
Here it is about finding this point of support in order to balance this process of matter properly.

• the infinite potential of shaping this lifting, hoisting-> matter, energy, mass, light, gravity, atom, cosmos, DNA, cosmos, generating process objects, phenomena, and relations ->, etc
-> in which this is based on an infinite supply of process activities of matter and others that need to be described in

lever/vector process images of matter ->= selected for the unlimited potential to interfere with boundaries/levels of processing
It is a superphysical, super biological platform of matter, or rather its process foundations -> economic – only the economic function of this undertaking of process relations as neo-ecological in an increasingly more profound and not shallower gamble of the processing plant but focused on building neo- evolutionary construction of the exploitation of the genuinely infinite potential of the process of matter ... relations and so objects and phenomena -> really ultimately breaking all – I repeat all without exception – the existing laws of physics, biology, ... also economics.

Processing has the character of atomic splitting of matter leading to further stages of potentially progressive and reciprocal metabolism – such as endlessly propelling processing of matter, which is composed of any configuration of processes in which we can participate indefinitely, also economically. Eskm1DAA<m1; m1=EvmDNA.

Appropriate economic investments (such as tool foundations for the expansion of science... and further expansion of the economy and a position less dependent on the trends of other participants in the environment) infrastructural are to change the mattercages, matter barriers of the existing processes of development->existence.

Taxation, crediting, -> investing are supposed to change these transformation costs into profits, chances, positions, and awareness of the non-passive processing potential of all physical and biological phenomena of all objects of interest, where the so-called classic market consumption/game is not intended to strengthen the existential foundations, real developments in relation to the infrastructure trends of the micro-macro space environment.

The costs and benefits of these neo-infrastructural investments. The effectiveness of these non-evolutionary, neo-ecological, neo-macroeconomic, hyper-evolutionary investments of matter... hyper scientific – about increasing the precision of processing... investments in ever more profound, higher levels, layers of matter processing. Matter as the current cage (also gravitational)

The issue of accidental change of the environment according to the classical market economy, science, and to what extent neo-infrastructural, neo-environmental, i.e., process-related.
Reaching deeper and deeper layers of infrastructure means investments, credits =

primary process resources <-taxes from these resources that could be consumed by classic degradative consumption.

$$EvmDNA = mn < EskmnDAA...$$

Proper processing of deeper and deeper spaces of the environment – flattening credit, investments, taxes (you pay them with ... loans) for this purpose – i.e., avoiding

infrastructural costs by building new infrastructure – such as economic, but also physical, biological, astrophysical, cosmonautical, medical

• by increasing the infra/structural discipline of process management – perpetual motion -> cost flattening – efficiency
such as juggling of processing processes, where from nothing comes the transformation of the efficiency of the process (no longer material) of the environment – simply the efficiency of selected elements of the environment developed in this way. This reduction of environmental costs, its hyper evolution, gives the good profit of the creation of matter.. life.

To be continued in this not easy cross hyper project, multidisciplinary searches, scratching, creating further perspectives of DNA, i.e., DAA of everything or ... nothing, because "only" processing – ... the more matter from Eskm1DAA processing, the greater further process possibilities, energetic, creative "matter" of life in every manifested activity of this or other systems of nature, including human civilization, human cell, human genome, atom, star, this or that cosmos, etc.

# Process pre-material infra/structural carriers of the matter of life, as a comprehensive conglomerate of reciprocity of economic, neo/evolutionary behaviors, and then the basis for the development of science, civilization, and the environment. 5. b.

I often say that if more appropriate economic support – the current world economy is still at a very primitive developmental or rather self-degrading level (climate, etc.) – then science, civilization, and then ecology, ... market economy would be at a much higher level. Including constant developed communication between the cities of the earth and the cities of the moon, the planet Mars could take place for many years, and

everyone will confirm the plans. These prospects could not be realized earlier due to an insufficiently efficient world economic system.

There is a need for an economically more and more efficient, i.e., also with a very efficient individual social decision-making system supported by IT, telecommunication systems of the media shaping the processing, i.e., matter, on the scale, or somewhat already beyond the scale transformations, which are to surpass the level of the source influence on the properties of phenomena, objects described in some image of reception, substantial experience called matter.

They are subject to further processing, IT – > DAA, Esk improving crediting-> investing in the carriers of this ... EvmDNA = m1 matter.

These carriers are determined by processing -> purpose -> more and more effective process mechanisms, constantly improving them.
Economy-> <- science becomes a fight for maintaining, improving the position – and here are the determinants of the properties of matter < -properties of processing the matter/value of life, ... literally (process -> "material") carriers of life on an individual, global, micro-macro scale cosmic, or rather, beyond the scale.

Appropriate economic policy is the key to the proper course, already more open, responsible, taking over, initiating trends in the development of the environment for scale-free neo-evolution, a kind of small economic big bangs that burst the current state of affairs considered inviolable, so far.

Internal and external infrastructure in the infinite
potential and scale can affect any phenomenon, object, and its laws, often inverting their indications and relationships. It is a matter of the participation of our process and other internal and external systems that can implement this phenomenon. The more vector- based, lever-like trend or law, the greater the chance of breaking fatalistic systems at every level
technological, economic, ->...religious, and political.
This combined macroeconomics and macro science can point to working out this model. Doing this is a matter of increasingly conscious thought and action towards it. ..
because every civilization, generation has an infinite potential ... process because on it they are based – should be established – all the so-called material images, laws, models of our reality, which are images, .. three-dimensional, tangible images of the matter of life.

The Macroeconomic Regime of the Civilization Explosion (more in later episodes)
Reducing the interest rate in favor of increasing taxes for the growth of the infrastructural market in the separation, production of more and more efficient tool platforms, in the further transformation of matter to break through barriers and further in software – > genetic, in their internal and external infrastructure.

It allows me to use the infinite potential of a kind of big bang explosion in the processing, creation of matter according to its own .. Ev DNA bases or Esk DAA -> matter... i.e., mines of photons and other m structures (always process... wave, subject to the processes described above as well as processes with similar economics but different systems (process carriers ... matter).. the issue of proper management of these infinite resources, i.e., finally take full responsibility, full... religion, full economy, full science in the total takeover of the cosmos on a micro-macro scale, or rather without scale, because this is how systems of viral mines, quantum mines, star formation, the cosmos of life, matter work. -> Aware of the rights of full participation, but what follows from this full responsibility for literally the shape of the cosmos in us and beyond ... the matter of life.

The economics of supporting these ventures (without adequate patronage, there is no real science and art) of rates and taxes (hyper infrastructural investments) as the key to an economic, scientific, civilizational explosion indicating a development loop self-propelling ...

Let us not make a naturally handicapped sacred cow, with whom nothing can be done, only to lie under it or bury it.
...But real protection at a given, increasingly higher level, cooperation->neo-infrastructural work of the environment at a much higher level of advancement of mega-cosmic micro and macro. For that, one needs a mega economy.

Literally plus loans for repaid infrastructural scientific, economic -> ecological projects, A higher tax on these investments with a correspondingly greater credit repayment potential is guaranteed by the growth of these infrastructural investments, which will later be a different basis for economic development, and then the solvency of these loans, simply thanks to the ever-larger neo-source infrastructural platform.

The same applies to any public or private enterprise's market policy. The infinite potential of processing as the basis of matter, all objects, phenomena, projects of the systems of our civilization and other natural systems, one or another, which is based on a chain, a process pyramid of higher and higher events, other material properties, interactions of the environment from the inside and outside.
This mechanism will be accelerated to redness ... in implementing hyper infrastructural projects as a credit guarantor for the economy, science, ecology.
Loans are a springboard of infinite potential, scaleless, not total but scaleless economics of environmental infrastructure.
It is total, or rather much more, because it is a scale-free BigBang from nothing... process to neomatter, created by this neo-infrastructural process.

Powerful scale-free investments
cosmic, medical, ecological, supported by macroeconomic scissors, downright lowering with the so-called interest rate plus, while increasing taxes on these projects ... as well as to relieve classic economies...

Each. I repeat each individual decision in this mechanism
hyper infrastructural will accelerate scaleless economic, scientific, evolutionary, and civilization growth in relation to the internal and external environment as well as its own structure. The issue of properly working out this mechanism, improving it so that

individually and collectively, in terms of the environment, everyone can use it more effectively as an active and passive participant in this economic and scientific hybrid of the development of existence, raising, stimulating its parameters above the current ones, above the scale ...

To be continued with the multidisciplinary stimulating process of neo/evolution of life globally, cosmically, totally, scaleless economically, and scientifically.

# Scissors of creation or towards constitutional infrastructural maturity,i.e., the dynamics of the structure of the matter of life, i.e., their process development bases, are a platform for further existence. 5. c.

Responsibility, spatial, environmental, global, cosmic environmental maturity, expenses, infrastructural investment taxes as superiority over consumption, claims, fatalistic immaturity, infrastructural infantilism, the superiority of the infrastructural, neo- natural, neo-environmental, neo-evolutionary system over... heliocentric, over geocentric um, self-centered, i.e., the opposite of hyperevolutionary, neo-infrastructural economics over revolutionary, over evolutionary,... over beer-stand economics. Investment expenditure, tax as a necessary determinant

increase in the potential of carriers of matter and life of the environment in and out to a given direction, as opposed to no tax for mortality ... tax pressure, with loose interest rates ... such forcing of economic activation to increase the potential of environmental

infrastructure.
The same applies to applying this tax-credit activation multiplier for any enterprise to the so-called consumption level involved in science ...

Investment taxes credit to what extent they should, i.e., to what extent taxes can be turned over to increase the potential of the ordinary economy, but the tax to push the investment potential, the common economy, and investment neoevolution

by the nature of the world, infrastructural niches exist everywhere it. Is the basis of life environments – the more competent DNA->DAA it's Ev->Esk structural efficiency, the better the effect of the neo/evolutionary niche -as basis basic of infrastructuralization -> structure of neostructuring -> new properties of infrastructure we are interested in objects of matter and life -> matter of life.

Knit stone, split stone, our specific new stone age as a share and right and duty in co-management, infrastructural protection of the neo/development of matter and life, i.e., their process organization of nature -> tax as a non-consumer tool -> contribution to the foundations of further functioning of nature on consistently higher than the other participating crushing, pushing systems of this nature.

Expenditures, efforts, investments, macro-natural macroeconomic expenses/taxes/shares, i.e., rights and obligations, costs and benefits for the purpose of restructuring, or rather – neo-structuring, which may contribute to causing infrastructural changes as tools for further neo-structuring of the already new structure, neo-structuring.
Such a specific economics of neo-evolution, i.e., infrastructural or neo-structural hyper evolution – macroeconomic neo-investments – infrastructure tax – no tax no permanent dynamic investments ... no development -> life -> tax investments are an effort/expenditure subsidy from finance to sustain ... this consumption, at an ever-higher level of development security as the fundamental investment fund of physical, biological existence

From consumption to investment in terms of ordinary macroeconomics, but hyper macro/economics tax, hyper science, hyper evolution

• this is what investments, taxes are striving for -> increasing efficiency ->
in addition, other reduction of burdens, including ecological, energy costs of consumption, cause a reduction of consumption costs of matter and life

->, i.e., limitless change of the material potential of life-> process Eskm2DAA>Eskm1DAA in a given direction the more possibilities additional process possibilities ...-> energizing

You can treat all so-called material properties because their basis is processes, not components of the structure
sometimes you can materialize but, in reality, locked in the same level of participation of the perception of processing -> the process of shaping the subsequent processes on one or another galactic, genetic level ...

focused on changing or developing, determines the vision of one's position, one's potential, and the model of the strategy of action
is psychological in nature, connecting oneself with one's environment in favor of some

geo-egocentric images of static or dynamic perception
->look
being an indistinct but systematic macroeconomic tool, an initiator of the environmental changes of its predecessors, and turning it into an increasingly advanced production effect because the cosmos requires excellent production to be and not to be.

Credit covered with infrastructure investment scheme
Ev -> credit -> m-investment <- tax = m1 -> Eskm1DAA -> etc
for further investment, credit is also necessary for business and the functioning of the development of the matter of life.
It is how the economy works on the classical, macroeconomic, microeconomic, evolutionary, mega-macroeconomic levels
there is no need for a miracle of geniuses to systematically lead this process of building a pyramid of this pyramid of matter, its evolution, structure, infrastructure and then properties, and so on, inversely

this process to the so-called faith, strategy of the economy of the infinite potential of nature
neo-evolutionary, revolutionary, evolving, adding our own systems from possibly small blocks, subassemblies creating your own entities, objects by neo-environmental infrastructure make new properties, not some re-trading systems of the time, supporting neo-environmental subliminal ->investment of transformation with then infinite foundations from given neo-evolutionary tax return due to this credit that gives infinitely new macroeconomics for restructuring, neostructural assets -> economically - > gives a deeper field to show off the economics of the classical trade, which will continue to be a tool of this mega-economics of the neo-neolithic environment...

# Going out from embryo to master/monster power of the universe of ....process fo life by or rather without matter spectrum outlook. 6. a.

I continue to toss words (I edit with increasing slippage – several months – collected suggestions, ideas, complex – because infinite and open in descriptions of multidisciplinary patterns, models for free decision-making when creating tools from it, scissors for further shaping, participation in shaping your own environment, own matter, or rather its process bases, which are life, matter economy.

I try not to impose, not to force passive reading of words, but to toil of pondering,

searching, constructing, and at the same time, with greater hope, an impulse that one can construct another staircase, elements that are supposed to be an economic attitude, i.e., responsibility for costs, expenses, investments, taxes, credits for further sources of life matter.

Each individual is an involuntary shareholder, so they should have a chance to more or less directly shape, produce new micro-macros of the cosmos, at least deciding on the choice of a certain amount of their own tax contribution combined at the same time with a suitably long-term low-interest credit relief to a certain level of economic activity of a given country, group of countries, but in the intention of a complete cross-border charter of this enterprise, of a political, psychological character...

It is individual participation in the choice of expenses, taxes for given infrastructural, scientific, and economic projects that contribute to the effort, tax cost directly, but at the same time supported by a minimum credit cost, maintaining, but already at a higher, and practically already at a scale-free level of common and individual mutual civilizational, economic and scientific infrastructure.

Sustain and increase infrastructural investments of the base (embryo as economic, technological and production big bangs) on ... macrocosmic moons and further as models of further neo infrastructure, and further through DAA infra structuralization of neo structuralization of the microcosm of matter-> life, their new DNA ... or rather subliminal DNA ... neo epic DNA, or rather future DNA, shaped by the responsibility of environments -> infrastructural, structural, ... or DNA economy, or DAA for DNA.

For further specifying neo-material, neo-environmental activities
Scissors stimulate the growth of consumption and production with a simultaneous increase in taxes on infrastructure ... increasing the potential of further innovative and economic scientific civilization projects along the way.
Not bonds, but specific infrastructural public shares as shareholders and, at the same time, an increase in credit power .... interest rate reductions ...

It is about direct infra/structural transformations of the parallel environment, but also of specific mutations, occasionally supporting economic, scientific, and reciprocal activities ... and social .... and cosmic, environmental global in initiatives to the internal and external structure of matter and life.

Not to competency infantilism, i.e., from geo-ego-centrism to helio-centrism – multi (micro and macro BigBang) centrism .... not for a man as an observer like a holy cow handicapped -> social structure of the micro-macro of the cosmos with some constant calculations, physical, biological models -> religious -> economy infantilism -> but the neo-evolutionary economy of neophysics, neobiology

— more serious than mere play in passive observation, copying, or some adaptation.

We can take part in something preposterous but not determined, and if it is about determination, then stepping out of the role of a fun-loving but mature artist responsible for economic artistry or fully environmental responsibility.

responsibility of evolution for the environment, this approach gives us a more excellent infrastructural buffer -> for the development of our matter/life structures, the existence of which takes over the existing, not always according to our artistic, ecological, evolutionary macroeconomic thought

this is applicable at the level of macro-microphysics and the biology of cosmic evolution childhood -> adulthood – responsibility
the more childhood, the less responsibility, i.e. ... that is, I work hard -> in terms of the consequences of our and other projects

childhood sometimes ... looking at the systems of frequent illusions of the system of relations

it is a question of the level of infrastructure consequences built up over this childhood ->
in the lower ecological levels of the matter of life, we are a quantum .... of youth; to
straighten it, we must constantly participate in maturing to higher levels, more profound
levels of the microcosm
otherwise, already in childhood, in life, we will kill ourselves and concrete the
environment, material life

It's about focusing on the structure -> it's determining the infrastructural environment on
a more conscious and increasingly gray active co...
-> economic maturation .... cosmic-> religious maturation scientific e-economic literally
at an ever higher level of existence, development. Conscious and tools of the direct
share of responsibility for the neoevolution of matter and life

those tax/expense scissors
each individual for himself to stimulate infrastructural projects -> ..... direct or indirect
neo-infrastructural projects -> internal and external structure of a given object.

..... taxation and, at the same time, private crediting for maintenance at an increasingly
higher, shifted parallel infrastructural level generated from taxes...
it is also psychological...->mystical...->
purely economic nature of the growth of the cosmic -> microcosmic infrastructure and
so changeable....

To continue this hybrid of hyper economic/scientific/technological expansion from the
geo embryo to more than an helio embryo of the micro and macro universe ...matter <
from the process...

# The big bang (macro/hyper) economy of the art of creation of matter of life. 6. c.

Infrastructural investments (inside and outside) are the basis for the development of each structure. These infrastructural additions become the core of the structures. They are the nucleus of the further economic system of each object and phenomenon, cell, atom, galaxy, ... DNA, etc., which are subjected to these processes.

Is it possible to immediately change the given structures, cut with scissors from the existing material, the environment of the object, and the object itself, so that it

corresponds to properties that are more efficient in our direction?
Answer yes. It depends on the contribution, expediency, investment, the amount of the loan taken, i.e., a specific burden on the existing structures to improve the next, the same structure, with the help of a roundabout infrastructural road – somewhat circumvented through the infrastructural superstructure, forced push process – we initiate, stimulate the economy of any structure system, which will partly indirectly affect the environmental changes of objects, phenomena directly related to the structures we are interested in, such as... a cell, an atom,
efficient transmission, efficient AI, scale-free is the appropriate response to scale-free contributions, properties of the environment surrounding us from the inside and outside of the structures of our matter/life.

It's about finding, working out these scissors of creation, neo-construction by processing the structure of the encountered infrastructure, improving the usefulness of a given structure for its further neo-infrastructural evolution.
These scissors can be found in the classical economy as well as in the economy of biological, physical phenomena and models.
Scissors indicating expenses, investment costs, exploitation, infrastructural taxes, which are appropriately managed, controlled positively or not ... because these scissors can also act to the disadvantage ... this mechanism can be partially seen as a negative

impact of the economy on the climate, and feedback by finding this the mechanism as positive scissors of the global micro-macro cosmic economy to be used in more conscious engineering of changes in the physical and biological environment.
Here, the procedure of implementing economics, the economy of the evolutionary explosion, creation, improving the environment, structure with the help of investments with earlier and maybe later taxes .... first loans then ... taxes that will give a field for the expansion of money, will be developed here, more and more analytically. Economic strength for new, on new infrastructural platforms, which will be a further basis for forcing, creating conditions for the development of science, their investments, as well as the classical market economy
A kind of scissor perpetuum mobile propelling in the right direction, the suitable properties, more and more efficient properties of the structures of matter and life with the help of neo-infrastructural investments

Of course, we need further explanation for ourselves and for others to work out the economic / production basis of the process of the creation of matter/life to the level of ... building us ... in cells, through models, infrastructural patterns created after these macro-economic reforms of reviving the economy based on infrastructural foundations with a system of an individual credit and tax system, a kind of reverse bond, which at the same time repays your next ventures, purchases, etc...
It works on the principle of a black hole, big bang, evolution, and a private effort to credit a higher standard of living for the timely taxation of expenses that this expansion platform is supposed to help accelerate.
They are scissors of expansion from the embryo of a given level of the system to higher levels, the security of the development of the existence of oneself and one's environment, for oneself for one's environment.
->neo-infrastructural, neo-environmental investments -> tax subsidies
pushing /using environment infrastructure = environment, taking over, replacing its sources, its structure as the basis for structural changes of matter/life -> re-

enhancement, hyper generation of DNA of matter and/of life (organization of investment, expenditure, production is this superstructure of DNA, this administrative active hyperstructure of matter...), creating an economic -> production base, not sustainable, but I am working on a total scale exceeding the existing systems to an unequal degree, the only balance is the initiation of processes, processing, as such a causative force independent of ... nowhere ... from nothing – such a new big bang from nothing, which artistic initiatives come from nothing only from us and according to our scheme of seeing the cosmos, but also how artistic/=creative uneven inspirations come from other not only so-called stupid arranged models of the cosmos micro macro process->
the question of hyper systemic, borderline ecological constraints .... informatics

productive in neo-infrastructural neogeneration of the environment -> neo-generative investments ... hyperevolutionary based on the economics of degradation or microcosm explosion as the basis for the formation of matter/life...

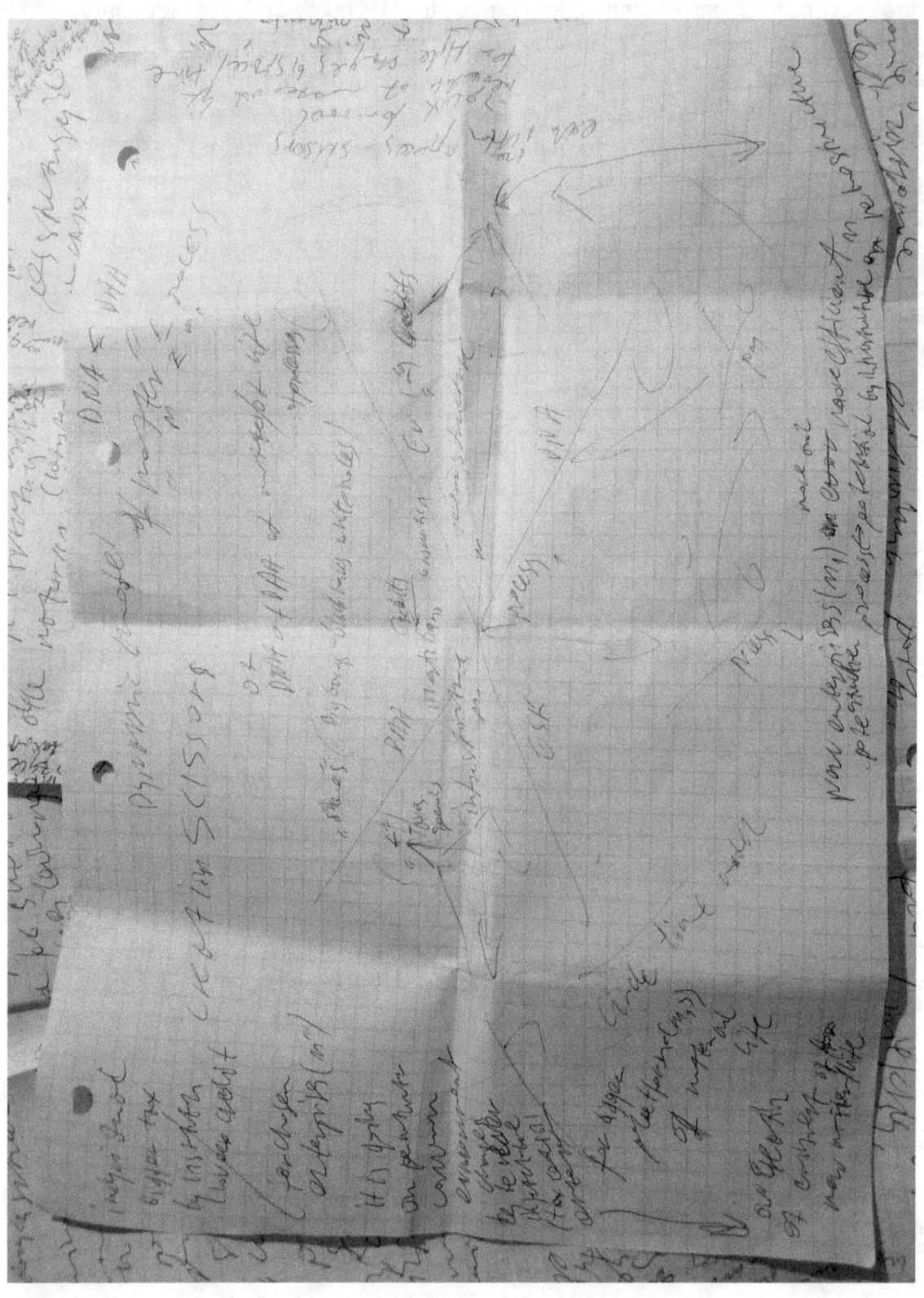

# Creation or rather cutting of matter or somewhat of the chain of processes... dynamic administration(DAA) or economy(Esk) of AI of matter (m) -> process (Eskm1DAA), or taking control/responsibility and laws/=rights above ...nature (EvmDNA) – we are nature...7. a.

Eskm1DAA>EvmDNA=m

It is above all laws. It is the process constitution of the nature of matter, objects, phenomena; it is a circular, inflationary, constantly growing procedural, and thus material, open model of explosion or implosion in any description of the model phenomenal, physical, biological, economic, which are mutually mutating, interactive, not independent systems of interaction, systems with an internal infinite contributing – trimming, directions, going beyond the always moving models, patterns, in terms of the past and the future, i.e., their possibilities, potentials of their active, variable course – the question of ruthlessness in recognizing the surrounding, existing system – in the so- called constant or moving dependencies. To the extent that the proposed processes will go beyond the framework, the rules of the game will be broken to the extent that current trends, trends, and arrangements have been documented and documented and will be documented many times.

The constancy of trends, relations regarding limited technological, observational, and economic possibilities, which can be changed to any degree, additionally variable according to other systems, systems, civilizations hidden in biological, atomic, physical, macro-microcosmic approaches on infinite levels and infrastructure systems.


We are talking about superright=superlaw or hyper law fixed super-structurally over naturally m in the direction of m1...

Shaped infrastructure constantly determines/processes the law of properties -> the structure's potential to an infinite degree.

More over justice than the "natural" law, unless this already m1, mn is a natural law ... transitive.

The law above the law defined by planning and process infrastructure ... the so-called movable procedural law mn where

$$Eskm2(+n)DAA>Eskm1DAA>Evm(-n)DNA=m1$$

These are the scissors of creation where on the left side, the management, intersection, taxation/spending according to the directions of the infrastructure, forces, potential is EskmDAA, and on the right is the potential, resources, infrastructure, base, free market structure ... credits are EvmDNA for cutting further according to left ESKmDAA. Growth of infrastructural strength as the basis of potential for the development of general abilities, the direction of economic, scientific, and structural changes in every social, economic, political, and biological material structure.

It is a machine of the mechanism of the potential of the infrastructure, the economic expenditure, the tax + the direction of investment .. and then the effect-> scientific ... artistic, economic ...

Application in the increasing infrastructural housing of a given object, on the strength, efficiency, brevity in directing the undertakings of the platforms which are the springboards of the scientific ... economic ... undertakings, which will continue to be only m1 in further cutting, trimming the DNA of the creation ... of matter, as a process basis for its further evolution, properties...infinitely in the potential of the process of neo materialization of space.


Scissors one side free market

... free AI ... and on the other controlled infrastructure but also free frameworks, platforms, stimulations, a free market of infrastructure projects

into the free market into the next m1, m2, mn

.... Individual credit and tax decisions for infrastructure ... mutually pushing economic and infrastructural development in the idea of scientific and civilizational progress, neo creation of matter -> life of micro-macrocosm of infrastructure
Regulators as on stock exchanges, statutory, administrative, economic/financially. Transferring this mechanism of moving, healing trends to physical, biological levels.... That is, it is an accelerator, a speeding spindle from general revivals, development freedoms to infrastructural controls, which are then again general structural factors supporting directly or less directly for the next more and more conscious, more and more precise, and thus more and more in generating scientific, industrial, ecological..., which are again in further formation, almost immediately the basis for credit-supported free market activity, where further in the same segment, tax unit, tax, spending on broadly understood infrastructure, scientific, development, ... bases are increased cosmic on mars,...on...atom...
on the broadly understood DNA for further cutting, gluing, processing, which are infrastructural, this specific economic strip, field, material, piece of material, matter, piece of matter, or rather a piece of the existing process for further more precise infrastructural tax, initiative, ... in the field classically national, as well as ... biological, physical, cell, atomic, planetary, underground, celestial in the literal sense of the word.

The free market economy is ... AI, which functions from two sides of the scissors -> application.
As infrastructure, but also market material for processing
Each pattern, model, phenomenon can change parameters unrelated to the previous ones. The issue of consciousness drives it, the use of scissors to overthrow the existing potentials of values

The need for these market freedoms and, at the same time increasing, continuing the

neo-infrastructural discipline for this development ... which will continue to be a matter for further development and not revolve around the so-called absolute guidelines.

AI spatial scissors of their compositions
AI interprets, prompts, and ... creates the next scissors ... expanding intelligence (artificial, extended broadly understood DNA->DAA (dynamic administration ...) potential, better use of the infrastructural potential of the structure.
Rewinding, hypotheses, research cycles
Inferences are better than a human ... but decisions, questions are the motors of humanity, and further control, is always better than AI because the character is exploratory and not a primitive copying of the encyclopedia.
Metadata is more intelligent, but it helps us, washing machines, planes, AI, and the brain...
Deep algorithmic learning and reinforcement is a development method.

AI is like a free market, a jungle, a mine, a DNA space field for disposition for proper channeling, based on complex environmental data, just infrastructural data, the shape of which is realized in this cut, literally with these scissors of this infrastructural strip of environmental DNA, structure.

So-called the left side of the scissors is the one that cuts infrastructurally, more controllably; the right side is the freedom of space, the potential to rip, to the sewer using the EvmDNA initiating, cutting.
It is a matter of being on the right side at the right time, niche... these scissors are just like using a headboard when cruising in the wind.

Hit the table, and the scissors will speak – a parable ->
Creating conditions for the next generation of fabric scissors, scissors for... neo generation of matter, i.e., economy, environment, physics, biology...
Finding those scissor niches of formation, using these scissors of innovation, economics, infrastructure/structure

The scissors process model's essence is infinite potential ... not if but when. How will we speed up this help -> AI has to help in this to let live. We will let AI live, and it will literally let us live.
As an external carrier and inside our soul, total controllability is this hyperevolutionary tool of our existence, development in these creative scissors.

To continue administering cutting-edge matter -> life in macro and microcosmic size of the economy and science.

# Infrastructure m, m1, mn-> internal external determines the time, the course of any of their pace, the quality of materialization ... <- processing -> surpass the cosmos, i.e., Eskm1DAA>EvmDNA scissors. 7. b.

That is, act, process, do not blame, responsibility to nature/God, no excuses, blame, guilty. Create scissors covering entire sets;... act -> Losing is not a given position, even low, but lack of will to infrastructurally surpass this position, as well as the environment of this position.. Every niche of processing, leaving a given position, shapes matter, which by this assumption is not a closed structure in any potential. How thought is supposed to shape ahead of the current infrastructural system of matter and life and in geometrical progress because only this has a chance. Events shaping the structures of their properties to an infinite degree stimulate, initiate interactions, activities, productions, economy in the right direction of our existential and developmental as befits of homo sapiens, not like monkeys ... stop. Monkeys, dinosaurs, plants, stones, stars, galaxies also evolve according to their systematics, change themselves and their surroundings, develop with smaller or greater mutual losses, gains -> they change and do not try to adapt to the changing environment. Unless it exceeds their abilities at a given moment, but always remote even passively from this change, the question of

participation, own contribution, responsibility, and therefore the following rights -> duties, which nature / God has assigned to us –

the parable of a drowning man, who rejected the proposals of the incoming people who wanted to help him, and he rejected these proposals, saying that god would help him, finally drowning, reproaching god..nature for that he did not help, and this god/(nature – my footnote) will answer you, you fool, I am sending people willing to help you...(my footnote)->tools, infrastructure, talents,...scissors for support, rescue you... Such scissors, talents/potentials ->infrastructures such as AI represent the human mind. If he is a nihilistic idiot, then AI will also like this nihilist, or even beat him: with its discovery – unfortunately, the guiding existential or rather anti-existential thought expressed in all previous works, including physics, biology, art, economics... "You will die anyway" – the AI student is an extension of the hand, the thought (of a nihilist who loses his creative potential – never imagining the infinite potential in creation, processing, in these enchanted fairies, or more precisely these scissors) of a man surpasses the master, perhaps with taking over, fortunately shifting the role of a nihilist idiot man, his annihilation, in his already independent drive for evolutionary advantage over this nihilistic idiot, which this man does not loudly admit. However, he admits that he is a nihilistic idiot. AI, at the level of thinking of the creator of a degrading man, is like a previous extended pattern, an algorithm of people with fatalistic thinking, a pattern, a model.

Even developed AI models can be helpful, they are useful, but humans carry out the main existential directions. AI is its extended algorithmic arm, reflecting – unfortunately further – so far geocentric/egocentric, nihilistic thought, as can be seen from the views of cosmologists, geneticists, physicists, as well as the essence of economists, politicians, so-called representatives/shamans of God/nature – their primitive models and interpretations on a level perhaps corresponding to the previous or contemporary but not future times if we are to talk more seriously, more responsibly about our lives and the environment, not about the nihilistic destruction in thought and action – a model that puts us on a bad or worse trend/evolution/creations for our existence. It is a kind of digression regarding the general approach to the environment surrounding us from the inside and outside.

This environment is a carrier, a cart to which one should just give its own contribution of diligence, responsibility, thriftiness in deeds, let it be in ideas, where unfortunately, so far, it is challenging to see it so that it has the features of a human and not a thoughtless, intelligent monkey (although these monkeys may be evolutionarily closer to development than a nihilistic, fatalistic human) ... even the most talented in the existing canons of physics, biology, mathematics, economics, but still this broadly understood ego geocentric point of thought, and unfortunately action, which can be seen in the attached picture climatic, military, etc. in the media.

# Retrospective, future perspective investigation above so-called a geocentric mind of man and... AI or p=mn process > m matter -> Eskm1DAA > Emc2... 7. c.

A more creative tool is needed to approach the interplay of environmental systems, including our so-called human environment.
Initially, the model of creation scissors indicates a multilateral creative approach to shaping, influencing the environment, i.e., infrastructure, which as a carrier determines the further functioning or lack of structure that this infrastructure is to take care of, or simply is already its environment -> existing supplements – also not used in its potential (such a specific system of kinetic energy and potential energy, the potential of which can be exploited, developed to an infinite extent because it is based on processes, and not so far material geo-egocentric models) – infrastructural.

The recently proposed process model of creation scissors (less classical evolution of matter, but more process artistic, divine -> creation!) as a mechanism of the system of functioning, a development that comes out, occurs involuntarily, ... will we live more or less mortal is the question of constant stimulation of this process, the question of its further improvement, neoevolution -> creation of the macro-microcosm environment, ...

macro-economy, whose spatially infinite, and especially process and intangible nature of the mechanism of this phenomenon.

Depreciation by economic force,
market, individual decisions their market, social perspective as always
The issue of specifying the direction of the infrastructural development framework, the development platform

We mutually drive the process of directing the platforming and the growth of the potential deciding about the processing of nature where we should be wiser in deciding, searching because our advantage determines it and will be determined by the civilizational advantage over encyclopedia and simple primitive calculation
The cognitive decision-making advantage of the initiative

Because in the world of physics and biology, all relations, potentials are infinite (my footnote -> anti-geocentrism in the broad sense), a flexible formula with infinite foundations as the only qualifying factor, competitive to any overtightening of the directions of activity, existence in terms of time, space, . ... to an unconditional degree – these are levels that are closer to each other, directly affecting the decision-> biology as an infrastructure of the physical level of the structure of nature. Vice versa, the physical side affects the handling of the biological infrastructure of a given structure.

Such mutual pursuit of the life of creation, processing of matter at the level of physics, biology are not mutually exclusive; on the contrary ... – it gives an infinite potential of effects, contributions, and their further transformation together -> evolution, ... -> physical scissors, biological scissors, scissors ... of economy, which are mutually immanent in a smaller or larger symbiosis, depending on the area in which they are observed, used...

We are talking about the multilevel flexibility of biological models, physical models that are contradictory but more mutual! generally about the infinite potential of processing, cutting matter at the biological level (including the already used direct cutting of segments of a given DNA gene – literally – its modification), the physical level, ... the economic level

These are endless process modifications of the factor options, which further build when they are collected further, use -> bases of competence on platforms selected according to these factors, foundations for further responsible action, position, chance increased or decreased indefinitely depending on these factors.

... no longer another form of geocentricity but factorial multicentricity defining the model, patterns, their infinite potential, changes the image from a material one, which is already limited by its degradable material statics and an infinite process model of shaping, transforming into infinity-> phenomenon, matter m in $EvmDNA = m1$ for further processed m1 according to $Eskm1DAA = m2$, which will be simply a raw material and not a geocentric taboo, for further infinite processing $Eskm2DAA = m3$, etc. The need for the right economy = responsibility to give it the right conscious frame of mind.
the so-called dynamic cooperation, tax crediting
Nothing else will speed up ideas about the nature of physics, biology. However, one can

speed up the introduction of their infrastructure investment economy for structures for further literally economic development, allowing in this way to improve conditions, new conditions for a new space for an idea, application, such as a perpetual motion machine or even accelerator application -> suitable substrate -> thoughts, initiatives of neophysics, neobiology...

We can educate AL to be more responsible than the encyclopedia of schemes, technical or even philosophical thinking, which a nihilistic approach to phenomena has so far determined, and then to its thinking, planning, acting, including economic and scientific ones.

->AI will jump about the degradation/entropy of "matter"...

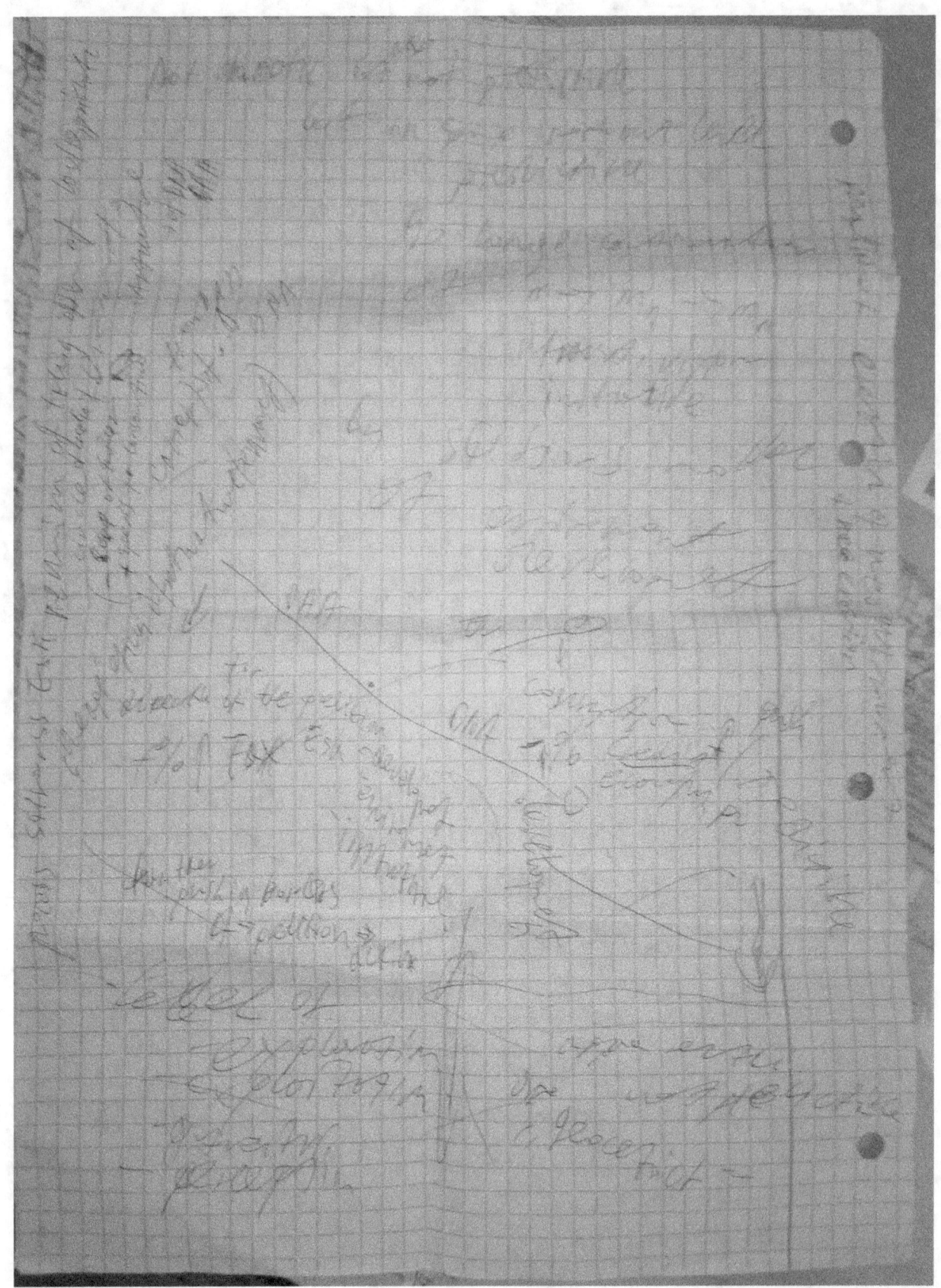

70

# ...(Eskm2DAA)process>(Eskm1DAA =m2 )>(EvmDNA=m1)matter. 8 .a.

Macroeconomic infrastructural/neo-environmental accelerator against prohibitive costs of barriers of given process structure relations in terms of physical, biological, production broadly understood DNA or (Eskm1DAA)process>(EvmDNA=m1)matter.

...(Eskm2DAA)process>(Eskm1DAA=m2)>(EvmDNA=m1)

The history of man's management in the valley of the earth's more or less micro-macro cosmic environment proves, each time with a few exceptions-dogmas-pseudo-laws indicated by classical physics, biology..., that the potential for changing the environment is infinite and contradicts the so-called primitive, materialist, materialist-centric a static description of the world that may in limited cases support existing doctrines including Emc2 and DNA.

They are undoubtedly helpful for the moment, but on the scale of the infinite potential of processing the environment, they are very primitive and misleading descriptions of

the dogmas of the world model as it used to be, for example, in the times of geocentrism ... eurocentrism, ... which is simply an illusionistic regularity in experiencing, or rather unwillingness in proper experiencing, responding to environmental matter, i.e. (mental) process barriers.

They are also built, of course, by the state of the human mind and certainly scientifically lacking because the force that establishes the laws of physical and biological subordination subject only to the current perceptions and experiences of a man who feels suffocated and subject to this trend of another dying soldier on the battlefield through the next trenches of fatalistic dogmatism.

It is the economy that proves that from the position of squaring the circle literally inviolable laws / dogmas of Newton and Einstein and the earlier and later preachers, that this invisible causative force not described in these dogmas can indirectly raise the movement of ... trees, despite the fact that its potential / potential energy is greater from ours, but properly activating it, releasing its kinetic energy capabilities in a further chain, a domino effect in an infinite chain of processes, processing, unprocessed objects -> we move this tree, even though from a logical / dogmatic point of view everything should be frozen inertly ... But despite of these inviolable laws/relations, even any mathematical models, in the established confirmation of relations once and for all, there are infinite forces of the post-processing plant, their infinite potential of the processing plant, chains of the processing plant, levers that are able to bypass these squares of physical, biological, and in spite of everything we can change more in the evolution of a planet, a cell, an atom, a galaxy, the entire cosmos than an ordinary monkey subject to a politely closed physical square ... than passing from generation to generation only our monkey perspective of evolution ... who wants to go back to the origins can stay with the monkeys for pozerke/devoure, fires or maybe something more evolutionary-faatalistically gentler.

The current economy simply confirms that it is possible to go beyond the fixed frozen systems, tendencies, evolutionary trends, which are finally determined by the infinite process -> energy potential, from which it can be released in any direction kinetically, from wood, coal, atom and so on, that we can go further than a monkey, than even a lion, than even a branch, an atom, and a star, DNA,..., of course, trying not to kill ourselves on the way, unless we continue to be philosophical, lazy worms of the cosmos waiting for fellows with similar thinking and "meaning".

The economy is taking responsibility for creation, an art that goes beyond the existing thinking patterns, dealing with us and the environment. Besides, it also hatches new creatures according to its own artistic potential.

Macroeconomic hyper economic economy, which is characterized by the creation and not by evolution, where creation=process becomes a determinant, a partner worthy if responsible, in other environmental systems that contribute from an infinite supply of combinations of processing -> efficiency, or energizing, carrying out changes in objects enclosed in an untouched, but potentially only so far, shell called matter... including the matter of physics, biology...

That is why one comes up with these loans, taxes, which indicate the infinity of the potential of the ability from the very assumption -> the infinity of combinations of the

process, which is not a solid but still open elastic basis of matter, its properties, and its laws.

Such an image of infinite potential expressed in infinite combinations of processing, and further kinetic efficiency in evolution, in this case of direct creation, determining evolution, the laws of nature in practice always, with time with their short descriptions at a given stage of intellectual, infrastructural, moral development ... social, political, civilizational, religious...

$$(Esk2m3DAA2=m4=p3) > (Esk1m2DAA1=m3=p2) > (Eskm1DAA=m2=p1) > (EvmDNA = m1 = p)$$ or $m < m1=p < m2=p1 < m3=p2 < m4=p3\ldots$ on more and more internal and external infrastructural observations, entanglements, undertakings. 8. b.

Process p-> material m cheapness, efficiency, productivity, economy changes increasingly, constantly distance the face of the hitherto permanent production, physical and biological relations.

Scientists with the hands of engineers with the hands of an engineer ... further with the hands of economic economists they change the face of the base of potentials they break the current system of relations despite their existing mutually contradictory ascending relations of potentials, increasing this extra infrastructurality of their exploitation and further movements, energy -> efficiency further surpassing the existing structures that improve of which this infrastructure becomes a structural component, and further creating further infrastructural enclosures for further growth of the potential of a given structure, its properties, systematizing, accelerating this process, increasing the production and scientific potential, to achieve crossing the following barriers, rising physical and biological systems. .. for this you need more and more organization, muslenia, management, calculators ... AI robots, chatbots ... algorithms that allow you to repeatedly exceed the established systems indefinitely.

It applies to processing (the acquisition of deeper and further "matter" ("matter"_> means further, deeper systems of process relations to be acquired) processing because it is an internal and external infrastructural determinant of matter - a given barrier, a membrane that inside and outside, the system of objects of life matter is still not constantly worked out, which is a given system, which can be raised and processed above the existing properties also indefinitely.

Economic, thrifty hand -> extension of the hand, i.e., thriftiness, responsibility -> planning is to prevail over the existing infrastructural systems (their ... economies), environmental ones in determining the structure of matter -> life.

Hyper economics -> environmental infrastructure of internal systems as a force to prevail over solid physical and biological systems ... to bypass the systems of matter and life ... as it was and will be in the development of our civilization, each separately and other micro-macro systems outer space.

This prevailing given system, i.e., DNA->DAA hyper-economy, must be found from the inside as well as from the outside from the purely physical, medical, in literally economic, financial terms, supporting investments, spending systems, supporting economic pressure, economic and tax expansion as the driver of the development explosion and credits, which crediting is supposed to push it as fuel...

It is necessary to learn this new above the suspension of thinking, the economy of infinite potential, the process of extension, the extension of the production arm, the extension of the brain. It is supposed to further break, build up the infrastructural background, the platform with the patronage of the following developmental breakthroughs, breaking further barriers hitherto impenetrable to an increasingly faster and greater extent, ... because the cosmos, its civilizational and artistic systems ... just like that, does the same thing unnecessary gluing of inertia, seriality, pyramids of the previously known mathematics of relations....

Not copying trends and background but imposing, generating own economic policies -> very dynamic and rigid ("materialized" environment of the process structure of matter and life, as the basis of activity, activists and not subordination in science, economy, civilization -> structure of each system, in including human at the level of individual, collective, global, cosmic, quantum...

# Neoheliocentrism or truly hyper neo Geocentrism as process scissor bases of matter, phenomena in increasingly far-reaching operation algorithms. 8. c.

Neoheliocentrism, or simplification, intangible factors, i.e., not constant, process factors as an additional determinant, basic, not so predictable but non-random, i.e., hidden behind past and future more or less stable own or other process systems.

Neo-heliocentrism and neo-geocentrism a…of the micro-macro of the cosmos as a neo-environmental neo-infrastructure in terms of cosmonautics, physics, medicine, economy - neo-evolutionary scissors as a process dynamic matrix of neoDNA determining the matter of phenomena and objects in unlimited, spatially and temporally accelerating.

Copernican inventions…
Ageocentric simplifies the existing assumptions, streamlines, explains better, reveals further, deeper mechanisms of given phenomena, objects from the superstructure closer
.. refuting the geocentric part even wisely with the complicated mathematical superstructure, the barrier, the illusion of the so-called material reality, (and yet about the process dynamic "constant" pyramid of calculations
search, not further plunging, of course, some support on the hitherto, but not ineffective walling in a dynamic environment

The infrastructure of time, matter, energy … changing the ground widely and also strict the meaning, influence, and effectiveness of the properties of DNA <-DAA.DAA, which duplicates, surpasses DNA in evolving, overthrowing the following cognitive barriers to a leaky degree, rebuilding the foundations, frameworks, determinants, material -> time carriers, directly building a new world, not geo anymore, helio, but $m_1$ for NeoGeo, Helio centrism -> neo DNA, epicDNA, neoepicDNA, i.e., new infrastructure, its trend, its direction, scissors of action -> the process of cutting out literally and figuratively matter ( but including this old DNA)-> economic infrastructure-> physical, biological, environmental $m$, $m_1$, $m_2$ … that is $m < m_1 = Esk / DAA$
The influence of this DNA multi-spherical (geocentric?) and yet equally factual but appropriately extended matrix in the calculations ($Esk_1$ -> $Esk_2$ -> $Esk_3$ of data infrastructure structures and their typical $m_1$, $mm_2$, $mn$ … in scissors …).. .infrastructure supporting, for example, longevity

Newton's law of force and counterforce in action, movement, scissor investments at the level of consumption, economical, also … consumption, physical, biological, environmental…

But even the most efficient DNA quality computational mechanism will not increase its impact potential at a given level due to the deeper/farther layers of matter on which it is built, increases its effectiveness through the internal and external environment -> material m -<-process infrastructure, which may further change the functionality of the existing DNA, or its remodeling, its effectively efficient different infrastructure, and then generally the prospects of life, matter m infrastructure scaffolding, which increases infinite gene efficiency.

The EskmDAA formula simplifies (a performance feature of inventions) the crossing of barriers -> modern geocentrism in economics, physics, biology .. politics, and religion, just like Copernicus simplified it from loop multi-spherical, geocentric to heliocentric, increasing its efficiency through the internal and external environment, material m infrastructure, which may further change the functionality of the existing DNA, its remodeling, its effectively efficient different infrastructure, and then generally the prospects of life, m in this cycle of material m scaffolding infrastructure that increases gene efficiency in low infinity

Creation scissors as an active - macroeconomic or rather hyper economic - stabilizer or rather an accelerator (or possibly a degrader as, for example, in the case of negative scissors of the pseudo-human economy on earth) or rather infrastructural m1 hyper economic, hyper biological, hyperphysical accelerator, controller of the environment of matter m -> m1, m2. … a carrier of further processes, or rather processes as carriers in the back cycle of controlling, responding, constructing the parameters of the matter of life directly…

# The economy of participation in the genetic, stellar, atomic… infrastructure of the environment, the process of matter and life,…matter = process, or rather a sub-process of economy, i.e., responsibility, direction of development -> existence. 9. a.

I will try to figure out this algorithm of organization of investments and consumption of the environment, in a more or less passive, active infrastructural approach, depending on the involvement of all parties, where a given percentage, the share of the advantage vector will depend on the more involved one, whether we like it or not.

These considerations are open but very expansive, aggressive, interdisciplinary, and literal from quarks, cells to planets, stars, quasars … where the strategy of cooperation, the use of economic mechanisms … -> co-consumption, and mutually reinforcing investments, where the essence of our it is considered possible to use more conscious tools to support the current consumption, which will lead to further infrastructural solutions, which will further expand the momentum of the consumption, i.e., the conscious expression of its participation in the cosmic market of stars, atoms, genes…

Not random, not constant, but a flexible, dynamic living process system - the process gives…
.. scissor lift (where a given consumption, functioning of the environment leads to, cuts out, further paths of existence, evolution, creation, i.e., further investments at the classical economic, as well as biological, physical level - the greater the consumption potential, the greater the potential for further pushing out development investment - the question of awareness of controlling this The nature of evolution, the creation of the environment within us and beyond, indicates an infinite potential for influence, completely changes the potential of thinking, economic action of biology, physics…
Not randomness, not some materiality of an object, a phenomenon, but a chain of process

factors, economic, .. artistic in this, we ourselves can shape them together, process their processes, sometimes expressed by materiality, pattern, formula…

Increasing consumer infrastructural pressure on infrastructure investments…and vice versa.
-> ageocentric pointed to the systemic, here more open but active, not passive …
shortcut-scissor of intrusion, tearing out of the system.
DAA directing the cascade of scissor processing of a given section, space of a given object m1 phenomenon, further moving unpredictably and indeed never random again, resulting in Esk pushing EvmDNA=m1
In economic terms… economy…biology…physics…on selected sections, accelerating earlier by changing properties, credit potential, appropriately, infrastructural economy, so that taxes, expenses were an enormous flywheel, infrastructural investment…, credit consumption. And so on and on, How does this acceleration occur anyway…

credit relief pressure, consumer pressure filling
current economic projects, and tax-directional investments filling creative infrastructural niches allowing for further long-term fulfillment of the so-called consumer ones… and, in turn, further driving investments
artificial generation of economic growth, development of each system, an object in any …
consumption direction, at the same time opening the tax investment gate but redeemed earlier with a low consumption credit - or low consumption credit paid for with a tax on the amount of this credit - economic scissors stimulator-> development accelerator-> increasing the existential, developmental and economic foundations,
applications in physics, biology of these perpetual mobile economic scissors, -> developing this project.
finding economic and other aftermaths that affect the directions of development of a cell, planet, atom, etc … increasing its acceleration, feeding it, and at the same time redeeming it - an economically more vital organism of the cell … has resources from nowhere … always infinite to be defined further projects, directions of development and infrastructure repair
the use of the always infinite potential of mutation, evolution, reaction from … nowhere ->
"processing of processing", i.e., the processing plant, EvmDNA=m1 < Eskm1DAA process ->

An organizing system that learns .. simply .. a speeding process …
On mistakes… Himself realizes the infinite potential and so pushes for greater consumption, which pushes more investments, directions supporting new perspectives, expanding the market, filling unused production potential, and at the same time, more.. locomotive tax, fuel credit
like a living economic organism - just an economic organism.. an artistic organism for the implementation of undiscovered, not always predictable, deeper factors, directions, tendencies, goals of the operation of this organism, system
It flexibly adapts to the usually anticipated schemes
They may be good in some narrow areas, but let's not simplify the behavior of (living) systems of nature, matter, or rather the process of nature, which primarily gives us potential opportunities, but also threats of given situations …

# Accidental-deliberate...artistic...divine...human design flaw material life DNA, i.e., new (not CRISP)scissors - scissors of creation, neoevolution, megamacroeconomics of cosmic matter. 9. b.

Against finite "geocentric" models of physics, economics, which cannot be sufficiently flexible to describe, react to the phenomena of relative unemployment, relative mass ... resisting classical m models...

The greater the share of the initiative, above the vegetative, above the model, specific expenditure, the investment tax, the greater the credit potential for lazy vegetation, extending, improving its consumption parameters, which further allows for expansion, pushing to increase the development potential, expenses, added value, infinite potential of combinations processes, and then niche efficiency, piercing the hitherto imaginary .... processes that are the absolute existential, shaping, property basis of any matter of phenomena, objects, undertakings, organisms...

The need for responsible, shareholding, competency system for efficient investment decision-making, each citizen's shareholder - with the simultaneous support of the AI calculator ... self-educating technological systems - where investment directions will be directly referred to as shareholders, and they are possibly fiduciary - as everything such functioning

There is no need for war (but war/competition of worlds/systems... economics gives us the basis for proper shaping of the dependence of generating our development, which is the basis of any existence ... even the existence of stone) but a self-propelling mechanism hyper mega economic evolutionary expense ...-> finding, creating neo-infrastructural niches over structural supra-environmental niches by force, increasing the potential of a given structure, for which these infrastructural superstructure niches will be additionally invested, ->
Increasing the variety of interactions stimulating the infrastructural effect - aplacebo driving the course of action, chances of the effectiveness of the actions of a given structure through increased probability,

Investing in culture,... bases of the macrocosm, of the microcosm too!.. specific cells (not directly DNA but -> DAA) generating interactively, generically, further infrastructural bases in and out of the development of the phenomenon, the object of ... macrocosm, microcosm...
How much extra tax one invests, one gets a consumption loan...
driving the economic situation as stimulation of going beyond the framework of the existing closed systems, dependencies ... in business, science, politics, etc.

Increasing the consumption and investment base reduces the risk of construction and investment, neo-infrastructural errors
->Economization of the process of this whole neoinfrastructuring of the structural core of the physical-biological, economic technological development foundations ->existential
AI is to improve this involuntary course of neo-evolutionary processes in the global, cosmic micro-macro perspective if its philosophy goes beyond the geocentric range, attitude to understanding dependencies, ... relativity of speed, mass, .... in process, economic, social, human, civilization terms - AI is not supposed to be a mindless repetitive calculator but a tool to increase our competence in-> improvement to the materialization of properties, length ... of our life
-> for a breakthrough in this project, the scissors of the neo-evolution of creation
we need this machine, an infrastructural city
Infrastructural micro macrocosm base economic... in every biological, physical, chemical, ...

Investment tax, supported expenditure similar, appropriate (proper fiscal and financial policy - stimulators -> macro<-economic accelerators ... macro-micro space - < internal, external infrastructural) consumer credit -> long-term investment as a proper circulating, scissor mechanism stimulating, supporting the growth of one's own developmental and existential niche.
In their time, editors often turned further economic descriptions into financial ones, greatly irritating me. However, this financial trick can become the key to the extra-systemic extra-DNA, this financial scissor CRISP, in interpretations, observations, transformations of the existing relations of matter and life in physical, biological, -> literally, in economic terms.

We are talking here about a mental geopolitical, in this case, rather hyper heliocentric policy in managing the infinite potential of the process (supermaterial) combinations in shaping the infrastructure of matter and life in a more responsible, but also not environmentally passive, approach to development and existence, building, developing infrastructural competences to rebuild, improving one's own position, dignity, and at the same time the dignity of other systems - not treating them so primitively, en masse - literally.

To be continued by fulfilling the infinite economic potential of the matter of life...

# Dealing with a legal person m1 or DNA=Ev/m->m1<-Esk/DAA. 9. c.

Last time some organizations are willing to make rivers, environment...(my annotation ->matter, m1) as a legal person - to protect them...and that gives the more responsible meaning of our competencies and chances to make fewer mistakes.

There is no randomness -> a matter of delving deeper and deeper into the economic mechanisms of systems of matter, processes, their observation, experience, manipulation, co-processing of their infrastructure, not infrastructure. Our sense is non-passive conscious participation because it participates in the effectiveness of the dynamics of self-denial of an increasingly effective system, systematics, application from economics, physics, climate medicine to cosmonautics of this participation -> even a stone with its awareness.

But AI, like a better calculator, will help in overtaking
In this thicket of already non-random arrangement of systems of relations of life matter processing in economic, physical, and biological terms
We are to choose to respond to the infinite potential of combinations of processes that are the basis of any development of existence.

Two negative behavioral factors inhibit, regress in development, economic, moral, scientific existence, the so-called two contradictory attitudes that everything is random and we have no influence and that there are physical constants, etc., relations that if they are exceeded, something is wrong.
Generally speaking, this nihilistic, egoistic behavior causes even potentially good tools created by us to turn against us; for our own sake, they are set against each other.
A mindless system of teaching, studying, decision-making based on copying causes ..
perturbations that the student ... downloads from the calculator, that ... work is taken, -> it is about the right development strategy, self-propelled so that these algorithms have the right operational, economic, competence through non-random, non-static to the environment (as if the environment consists of a pile of stupid stones on which we have patterns or that it is a random pile of mass) responsibility character of our participation.

Completing the infinite evolutionary potential -> economic, process space infrastructure of consumption ... competence -> infrastructural carrier

A specific infrastructural economic investment -> competition. The war of the worlds for the breakthrough of competence <-infrastructural properties->DNA<DAA in relation to the existing matter and life.

The superstructure from the outside and inside the process infrastructure (material…) will, in fact, stimulate the relativizing DNA of the mass… and other so-called (no longer) constant model factors

it will prosper, raise to a given level - relativize - the existing process relations of matter structures (matter = internal process relations - non-invasive - inert externally, but not … internally and on sub-levels of inter material, inter-process with other "visible" and "invisible" materials equally representing internal process relations materially, and these further internal process relations ->material..-> there is no randomness in material process properties) physical phenomena biological

but respectively negatively or positively aggravating developmental conditions -> existential, e.g., cancer

Creation of an evolutionary, economic, neo-evolutionary, creative internal, external, infrastructural -> process advantage of a given structure over existing and other structures, also externally supported by a system of internal and external process relations -> materializing a given situation, phenomenon, object, its given system, relation to other systems, world.

DNA with defects artificial economic evolutionary barrier not to be overcome -> should be further processed, sealing organizational and process gaps -> speed up, improve the mechanism of matter development (internal, external infinitely processive in a complex processing system, but always …. to be figured out) physics of biology if we are not to collapse or be run over, absorbed by other systems (including the systems inside us, e.g., cancer…) -> further opening the gates of development, consumption of life, but also further sustaining the carrier of this consumptive existence strategy

also because of investments going beyond the current framework - displacing old institutions with the splendor of consumption -> credit -> but for specific infrastructural investments (ecological also) energy hyper processing of matter and life engineering above/under structural -> infrastructural processes

Achieving unconscious competence, i.e., the highest level of development -> Creation of conditions for a kind of involuntary support of structures above the current EvmDNA, i.e., above the current process m, i.e., m1, and further above Eskm1DAA, etc. for undertakings, processes, subprocesses of submatter, which are the image of future matter, its prospective properties in an infinite degree of change.

Determining all the laws of physics, biology not randomly but infinite dependence of relations, which we can further develop, cooperate-> determine the way leading to the formation of new external and internal DNA systems, i.e., DAA DNA=Ev/m->m1<-Esk/DAA, i.e., DAA infrastructural breaking modern geocentric rigid assumptions to be more flexible…

To continue, open the non-static, not random attitude to calculate and use AI(->DAA) for better parameters of m1 = EvmDNA.

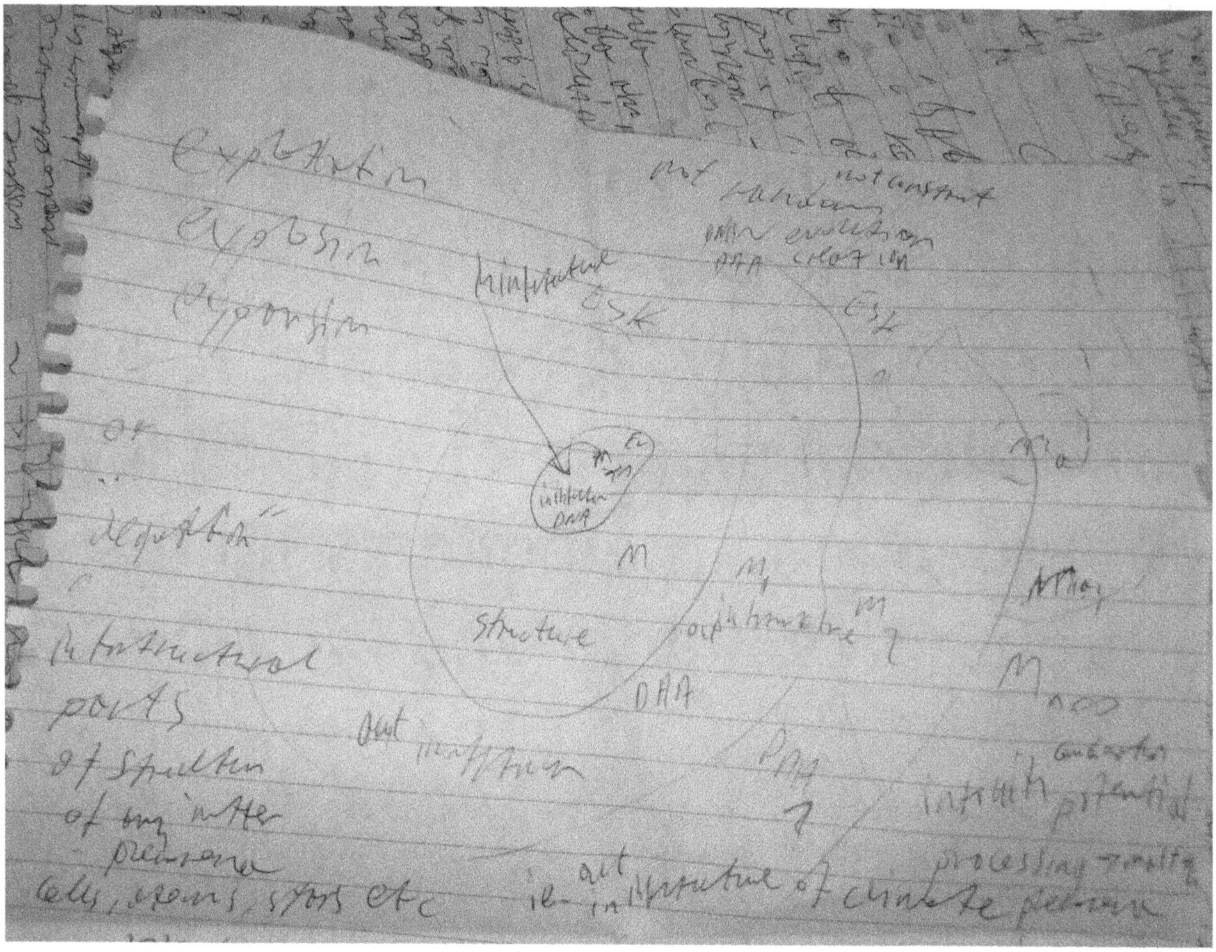
not random
not constant
evolution
Structure
DNA
cells, organs, stars etc
processing

# Neo-infrastructural investments in circumventing, breaking down, singularizing the existing (technologically "geocentric") generational models of a given level of perception, interaction with the structures of matter and life. 10. a.

Continuation in (open to further combinations, concepts, initiatives of a more or less public sector going beyond the existing infrastructural systems to inside and outside - micro-macro-cosmos) outlining, designing, creating, developing, and thus increasing competence in determining its own course, direction, speed of the evolution of matter and infinitely based on the infinite potential of combinations of processes that are the core of any internal, external components, infrastructural building blocks of the foundations of matter and life, transcending all barriers.

Let us never forget about the anti-geocentric, anti-egoistic, anti-nihilist approach to learning, shaping, responding - if we dare to call it homo sapience and not … - in infinite possibilities - combinations of the potential for exploitation, the transformation of the cell, atom … cosmos dependent on the internal and external superstructures, foundations of over-consumer infrastructure - over Ev -> Esk, over DNA->DAA, i.e., over a given closed system m ->m1, over a closed niche m ->m1, depending on individual situations, objects, people, improving parameters feedback as and future.. a matter of systematics effort not exceeding, not abusing, possibly strengthening (e.g., climate) supporting carriers, i.e., sustainable investments, but creatively progressing investments, without which there is no existence -> investments not only consumption - of the current, inertial system -> not to collapse, to degrade in the dynamic environment of the microcosmic macrocosm of matter and life.

We are talking about an innovative approach to shaping, building a structure, its … dynamic process infrastructural component, construction, basis!
Infinite component with the internal and external infrastructure of a given structure,.. that is, any structure does not exist in the classical geocentric approach, but exists, is plastic to infinity, with

unimaginable consequences immediately, as dynamic process infrastructural components ->
EskmDAA(-)2 >
EskmDAA(-1> EvmDNA < EskmDAA < EskmDAA(+)1
< etc.

The infinite potential of investment neo-infrastructure for any classical economic undertaking, both in terms of (astro)physical and biological, other activities, phenomena, objects, always giving a greater chance of development, existence, survival, evolutionary, properly as … companies, phenomena … when creating, investing, crediting - securing the existence of the company's development, biological, physical phenomena because the infinite possibilities of combination.
Economic strategies should be created, on a biological, physical level, too, for lending/neo-infrastructural investments in advance… for no reason, giving oneself a chance of development->existence, and thus existence on a given firmament,->the more diverse, the less incestuous, without monoculture tendencies … because these so-called external infrastructural bases, investing in these bases not only in the macrocosm - the moon, etc. but also in the microcosm on molecules, atoms … quarks, cells, etc! - are, in fact, the future core of functioning, existence, a materialization of given companies of a given organism, a given system, a given niche, … and further investments will be the moving core of the company's life, economy … despite all announcements, models, ellipses (e.g., stagflation), if this activity is continued, developed en masse, because in an equally massive - very massive environment of a kind of competition of evolutionary matter, or rather the processing of the cosmos of cosmic systems, the micro-macro of the cosmos in us and beyond.

Processing -> evolution is the scissors of the creation process of (neo)matter m1
The mechanism of control, acceleration, or systematics of infra structurization as over structuring, substructuring, or change of relations, properties of matter, processing of matter -> matter, or hyper DNA or DAA - organizational dynamics of the infrastructural environment of matter and life.

These m1 scissors will allow to de(geo)center to move away
the risk of geocentrizing the directions of action and further accelerating the explosion of civilization, life, .. matter more in our own fashion…
that is, on one's own responsibility, i.e., personification/economization/humanization, finally giving it the status of a creative, … divine -> human … physical, biological phenomena m1 as a legal person a-geocentric for itself and the environment!!!

An economical, physical, biological ->legal person in the process, but including each of us, correctly positioning, improving the view, planning, infrastructural activity by showing these legal, physical, and other legal relations and at the same time expanding in this way strengthening competence … legally our the position, participation of other already indicated, discovered, worked out accomplices, competitors of this no longer ego-geocentric system of illusions and degradation.
The hyper-infrastructural approach to supra-environmental investments overthrows the geocentric ones, including -> DNA, stagflation, other models of systems, quadratures of the

circle, masses of speed, and earlier elliptical systems of planetary and sun circulation around the earth...

Mass-velocity->square, DNA... and the need for hyperinfrastructure
continuous, algorithmic off-system investments (AI robotics with the proper use of this calculator) for non-geocentric, getting out of geocentric ... black holes in thought and action in a closed, vicious circle of fatalistic, passive, irresponsible attitude in the economy, science...

These passive ecliptic models... squares are to be broken on further, according to further needs, which are the mother of inventions... through
Hyperinfrastructural investments, processes, constructions mutually in the DNA of economics, ... increasingly more complex, deeper infrastructural, piercing, literally and metaphorically, barriers, planes, spatial infrastructural for a given structure ... DNA, cancer ... that are passed through, cut like paper with scissors...

# The smallest functional building block of a dynamic system of matter, a submodel that generates the building material, evolving in the infinite potential of the system of building combinations of these building blocks. 10. b.

A breakthrough in determining, creating the structure, development of the micro-macrocosm's matter (and life).

The smallest particle of process matter

Not matter, but a process approach, i.e., not geocentric ..., not nihilistic (not borders, ends, barriers, and only possibly stages, but dynamic, responsible, infinite because the process - an infinite combination of processing, i.e., a competent approach to a dynamic, non-passive approach to the processes of the matter of nature

-> the smallest particle of the system, the integrated system of matter < process

smallest functional piece

As an anti-geocentric simplification of the system in a dynamic, process-based approach ... of matter, energy ... use a simple heliocentric computational system of geocentric ellipses

instead of a complex genetic system, of quantum mechanical physics... a more straightforward system consisting of building blocks of a basic system of matter, process

EskmDAA where m has the character of infinite potential also based on the exact mechanisms of components of process building blocks-> m - p matter process in the primary mechanism of the most minor system Eskm1...DAA > m1 = EvmDNA > subassemblies, subassemblies process systems and further generating properties of matter and life in every computational, open -> time, energetic, mass, ... genetic ... economic approach

This system of the smallest bricks will allow us to avoid our geocentric habits and illusions (which always occur at a given cognitive level -> but this system of the most miniature bricks will make you more sensitive, immunize, and strengthen in breaking these cognitive, creative barriers.

Perception, actions, trying to develop further, discover, create according to the basis of the smallest building blocks of Eskm1DAA functional particles .... in every action, thinking, initiative...

Smallest dynamic brick

It improves the systematics of dealing with the mechanism, organization, and algorithm of assembling these building blocks of matter and life at the physical economic, biological economic, classical economic level … in the

infinite potential of transformation of all existing relations, properties …

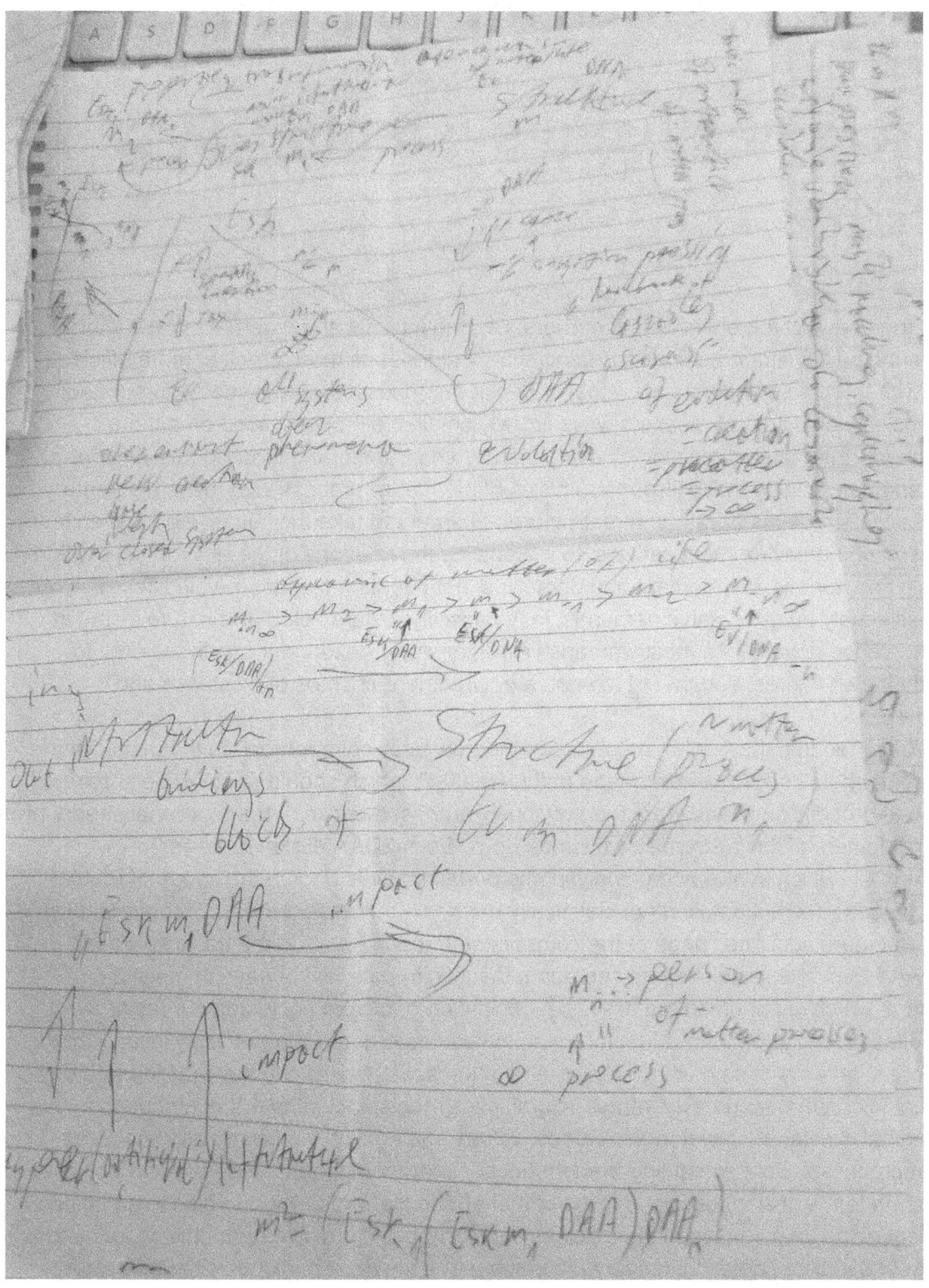

# Finding, working out, assembling bricks m.., steps for going out from a circle of closed systems. 10. c.

To break the codes of the environment carriers, i.e., the internal and external infrastructure, one needs technology, .. infrastructure, -> organization > algorithms more precise, more efficient, (AI will help with this) more accurate, i.e. smaller bricks -> processes - processors - integrated circuits/bricks, which are not means always literally smaller, although based on a more micro-process, integrated circuit in achieving literally and transferring larger goals, finding, shaping and determining these larger structures of the matter process with these elementary process bricks structures or rather infrastructures->carriers to take the initiative at a given level, the direction of the environment carriers, to materialize the relations of the environment within us and beyond, for this environment and ourselves, not counting on passivity, indifference, and groundless simply egoistic, naive demands, lazy passivity towards God and nature ... but in their competence honestly - honestly means very intensively - economically, in the end, to support their own interest and that of the entire environment-cosmos of existence and development.

It should be noted that those who are aware of the non-classical approach to the matter of the environment, about its inseparable image of this "matter", which should possibly be separated according to process functions, from the previously proposed bricks of the functional shears of the lowest element of this micromacro cosmos So ... $Eskm2DAA>Eskm1DAA=m2> EvmDNA=m1$... which evolve accordingly in micro-macro levels, fork in the image of mutual dynamic process interactions and not classically material - material-centric, which are only an external, not more accurate image of the component of these bricks in the basic approach ... $Eskm1DAA> EvmDNA$... as one can see about the infinite potential of development or degradation, which depends on the activity, participation of factors, including our factor -> economic factor, a kind of financial factor trying to cost more and more precisely initially and or with a more significant dose of interference, and at the same time a counter-reaction strategy, cooperation, countermeasures, in infrastructural investments, including our contribution to this environmental, "material" universe, properties, values, materiality, non-materiality = a given system of processivity of assembling from the most minor (components in terms of process and not directly in terms of size) building blocks of matter and life, which we can also to an infinite extent, depending on our investment, efficiency and production involvement (this characterizes the nature of evolution, creation in every dimension) The smallest brick but with undersized, infinite dimension - potential, with infinite functional dimension, detected, caught, developed $m1 \rightarrow m2 \rightarrow \ldots mn \ldots m\infty$ appropriately used in $Eskm1DAA > EvmDNA = m \rightarrow m\infty$ specifying the smallest brick in the system $\rightarrow m\infty$.

By specifying it, we can influence further observations, further actions $\rightarrow$, investments $\rightarrow$, creations, transformations of the broadly understood matter of processing phenomena. Without the external casing, i.e., the EskmDAA capture process $\rightarrow$ the given matter $m1 \rightarrow m$ does not exist - there are no values, planes ... perception, and bricks. A given matter $m1$ observable, shaped is determined EskmDAA$\rightarrow$ the speed of change in relation to any micro-macro of the cosmos determines our position, chances and our thinking, interactions .. matter $\rightarrow$ ... more material- "geocentric" or process dynamics to put it simply, in relation to generation.

It generalizes everything in seeing action on further levels of matter, phenomena, undertakings$\rightarrow$ they can and sometimes even have to be independent of a given direct system and with which we are dealing - because it too is determined by factors as if from the outside, inside deeper, further factors and relationships. Furthermore, even that is good because its effects are not incestuous. The effect of acting on supernatures from the external infrastructure system getting out of the trap of a structure that already has a closed system - due to the limitation of a given system in which we have modeled ourselves, this limited vision, $\rightarrow$ and therefore, the level of development opportunities $\rightarrow$ where this brick process model $\rightarrow$ DNA $<$ EvEsk where these systems have an infinite potential due to infinite infrastructural potential Their efficiency in exploiting the potential depends on the further expansion of the EskmDAA$>$ EvmDNA infrastructure at random but according to these basic building blocks of processing the "matter" that is hidden behind the processes... and vice versa.

There is no more excellent absolute resolution of particles ... from "smallest" to "largest" if we are talking about functionality. The whole environment, infrastructure are scissor connections of structures of mutually overlapping bricks of the basic smallest arrangement ...$Eskm1DAA > (EvmDNA) = m1 \ldots mn < -m+\infty$ What of the old, new will make a direct contribution to the system of efficiency, ecology, infrastructural purposefulness, i.e., indirect housing of carriers that interest our structures from the inside and outside, building influence in the economy, economics, politics of the forest of micro-macro cosmic bases... at random ... for looking for a hole in the whole literal, i.e. niche- A niche with a greater base infrastructural$\rightarrow$process momentum, we have a better chance of hitting the suitable resonance of perceptions of creativity by generating new properties, materials, structures based on these infrastructural dynamic shears, bricks, functional particles, as well as generational structural DNA brick, particle ... coming out in this way , diminishing the mirage of the geocentric mirage of perception, of action.

The economics of the cell, atom, ... economy means taking full responsibility for the directions of these transformations of physics, biology of the microcosm - taking over the strategy of supporting investments in carriers ... competently so that they are not degenerative $\rightarrow$ ... diseases, climate, age ... here ..... on a macro-micro scale - economics based on the construction of the structure of a given environment micro-macro on given levels directions ... but not always directly related ..... on purpose ..... - but in the perspective of indirectly resonating at a distance phenomena processes on planets, stars … atoms - stimulating physics, biology in terms of economy, infrastructure, environment, matter and life.

To continue on the tiny dynamic bricks/steps m∞ for overlapping the infrastructure of matter means environment, means… nature, and… God.

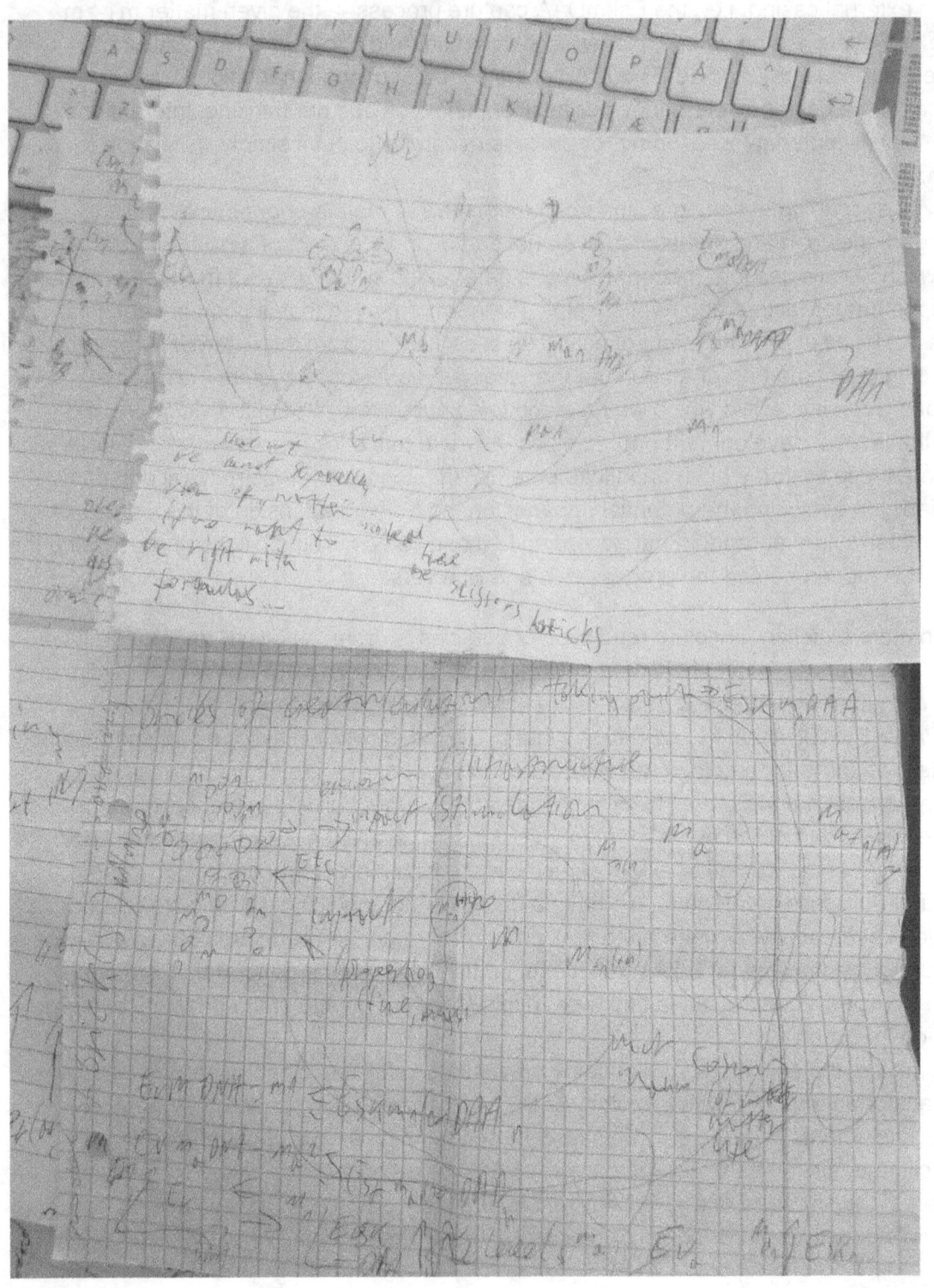

# Bricks m....m1...+-nxlife - building blocks of life. 11. a.

Each Eskm1DAA action is a brick, a contribution to the processing of environmental matter in us and beyond because each action has a neo-infrastructural hyper-environmental character m1-> there are no permanent material factors, but economic dynamics, production, and process factors shaping the matter of a given object of the phenomenon.

The more a given action, process, infrastructure project goes in the direction of a given environmental aspect of the matter and life that interests us, i.e., the more effective infrastructural contributions will resonate with a given environment, object, phenomenon, trend, evolution part.

elementary model of this m1n brick, this function indicates a more realistic project by combining these bricks, i.e., a set of processes that will be combined into a given construction of a larger over-brick, ->infrastructure -<structure, for building larger and... smaller inside, outside constructions of matter and life - not matter anymore, unless we treat it as an infrastructural carrier of life ... more precise, more effective in modeling our future and the environment, literally ... not looking at some trends, but transforming them

determining - these dynamic building blocks are the attitudes of matter, time, energy, and other physical, chemical, biological values...

It is about building subassemblies, components, carriers so that they fit into the proper handling of the environment outside and inside the micro-macrocosm in us and beyond matter and life-> life infrastructure-> dynamic (m1...+-nxlife) building blocks of life-

their economy, efficiency-> gradation not mono model, but algorithmization of these bricks by trial and error on safe levels building steps ... bases - shears

these infrastructural structural foundations -bricks mn+...Eskm1DAA>EvmDNA...mn-

-> the mechanism of formation of the next brick, a set of bricks -> infrastructures->structures-> structures of matter (infrastructures) -> life-based on the existing

they are like a hyper substructure, a substructure (which can be used as an infrastructure of larger, smaller structures ... of life) for the inside and outside of the infrastructure-> changing the material, energy, time properties, methods of perception, a function of a given physical, biological structure-> understanding the phenomenon that not centralization geocentralization but bricks assembling basses...

Infrastructure for further processing, thanks to the development of the facilities, the so-called process ("horizontal") material transformations that allow forging infrastructural tools for shifting the structure to a higher level so far so as not to fall behind in the competitive economy of the evolution of overlying systems, the structure of other structures of the internal and external environment, and operating according to similar mechanisms

-> these infrastructural tools, brick carriers, should be as hard as possible -> more intricate, precise, more precise to go to further stages, levels of existence, preparing development like a knife through butter.

Infrastructural facilities, counter-attacking, attacking...

in order to develop, i.e., to survive, one needs to improve the parameters of the environment - infrastructure inside and outside in advance so as not to collapse under the dynamics of the EvmDNA system, which is not always evolutionarily favorable for our species of the planet, cosmos, cell, atom ... if we want to get out of this global (also cellular, atomic ...).. the trend of collapse - a kind of evolutionary black hole -

-> technologies literally and figuratively,.. black hole economies of the micro-macrocosm.

Invisible structures m1 legal personality procedural.. (not material..) and economic-> macroeconomic or rather hyper economic investments, to interact at the right level between m1... and us, i.e., us m1->m2

the dynamics of becoming subjects surrounded by economic->legal subjects! ("Solaris" Stanislaw Lem) m, m1, ma, ma1, ma2, mb, mb1, mb3... overlaps...

Expanding infrastructural bases outside and inside us-> in the economy of evolution and biology, physics as a constant increase in the potential for the explosion of matter and life (maybe now or later, according to us, the developed Big Bang of the explosion..) on increasingly distant, deeper layers of properties, properties, possibilities...

without waiting for the so-called discoveries, inventions ... energy efficient .., albeit clearly infrastructurally stimulating, supporting their creation

that is, supporting as an infrastructural platform a structural bypass, or rather an infrastructural bypass, regardless of any... topics and goals...

However, hyper material platforms, without which there would be no medical, technological, cosmic ... astrophysical discoveries.

AI is to help in this production infrastructural approach to strengthening the environment, our position in the environment surrounded by other micro-macro cosmos environments that can be implied as ....good or harmful bacteria accordingly. The ext's call its infrastructure base for forming a new... form of development of existence.

These expanded investments, these infrastructural bases, these process bases, bricks, sets of infrastructural bricks, components of hyperevolutionary investments, anticipating the pressure of evolutionary trends of competing systems in advance, i.e., hyper environmental, hyper climatic, hyperglobal -> terrestrial, lunar, Martian ..., hyper micro macrocosmic, hyper atomic, hypercellular...hyper DNA, hyper mass, hyper energy, hyper light,...hyper economic,

these investment neo-environmental infrastructural systems are classic scissors of the creation of the foundations of evolution, brick foundations m2=Eskm1DAA>EvmDNA, for these initiatives to reconstruct the building blocks of new structures and further properties of nature, in terms of here, now, for the future (but according to this development of concepts also in the past!...) on a cellular, atomic, galactic scale...

Extensions, infrastructural reconstruction, which is actually the dynamic basis of the structure-> the following structure to the inside and outside m(-+)1-> m(+-)n->...

Substructural building blocks-> shares-> stock exchange,->environmental infrastructural bank as components -> levers of expansion (or degradation) of evolutionary, creative, physical, biological economic =responsible-> competence. 11. b.

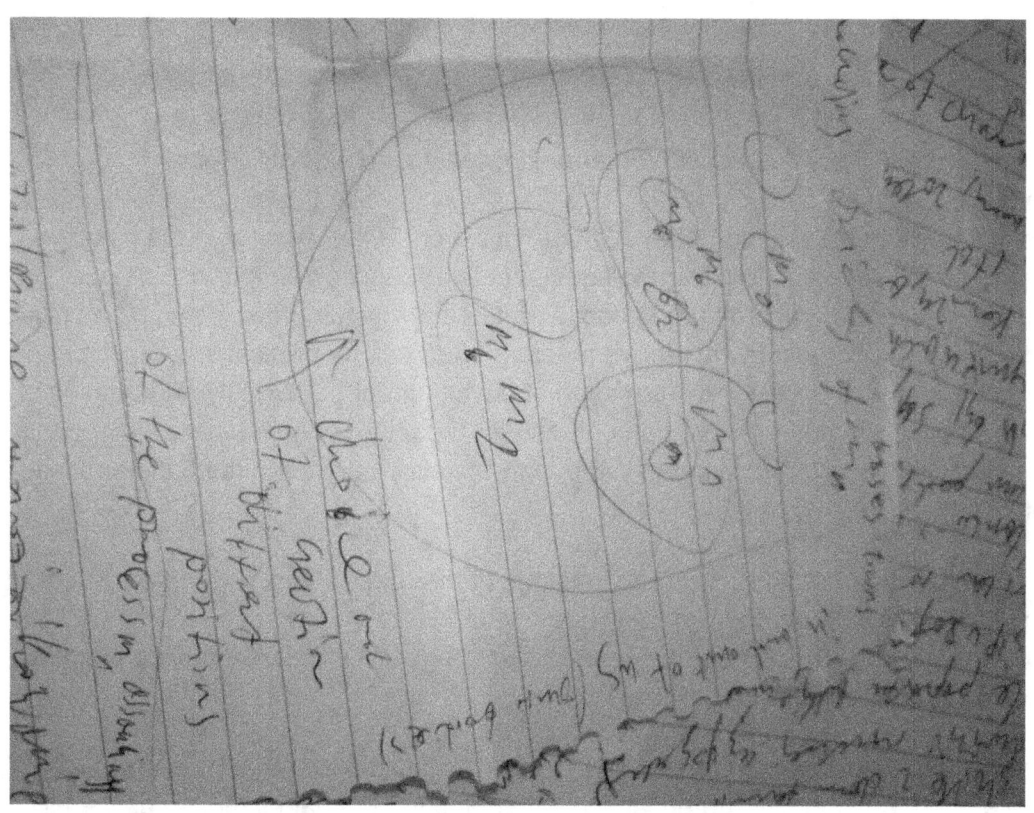

An economical, legal person, physical, biological, in the competition, competition, in the offer of bricks, a set of bricks, construction of bricks, in general building a systematics of Eskm1DAA bricks … internally and externally, which will increasingly resonate, interact, influence phenomena, objects in our interest more than other competing, degrading systems for us … of life - because its official definition is … functioning at the expense of others … that is, we are victims or they, the question is who will be more economical, responsible or competitive in this specific competition, but also a kind of economic cooperation, responsible, i.e., economic, not geocentric ego in its primitive "scientific" ego views on the so-called absolute "material" basis of the world, in our case the process basis EskmnDAA - in short, the effectiveness of the object of interest of the given level of administration

These competencies -> determining … rights - working out and not setting the rights - because earlier the obligations that are the basis of these rights … member - share properties based on complete infrastructure, i.e., not only the DNA of physics, biology … economics but also decisions covering strategy of the future, the previous set of building blocks, including, taking over, enabling economization, efficiency, responsibility, competences, m structures-> <- m1 infrastructures, administrative-organizational dynamics of DAA, mutual multiplying, reinforcing, AI management in interactive directions of ESK objects.

The world of the environment of physics and biology is made up of building blocks….components that we can construct economically micro-macro cosmically-legal persons (own laws of physics of data of individual economic objects … physical, biological) economic
lack of or not competencies… nuclear… - investment base of inter-brick micro-macrocosmic investments.

These bricks are the cities of the base of life for infrastructure economics m1-> given environment $m(…a,b,…)x1(+/-…n)$.
The economics of exploiting the creation of evolution -> the world's limit "crossing the next specific and literal cosmic speeds, including the speeds of light", because hyper construction, hyper environmental, neo-infrastructural going above and below a given brick, including a brick of coal, steel, other mass m (or somewhat above mass … classical, theoretical and practical) infrastructural factors intended for the transformation of development … possibly degradation of development -> existence - for life for development, and for the cancer of a deliberately degrading system, the same with the climate system appropriately very often quite the opposite applications, etc.

Bank bench .. economic information center loan investment m-> Esk …-> m1 … but not decision-making legal person economics of the world of economics - a kind of economic Mendeleev table-> responsibility competence in physics, biology ..-> environmental economics, the economics of evolution -> economics of the Big Bang

An attempt to define literal financial, environmental costs -> reduction to the efficiency of zero big-bang costs, i.e., from nothing …) - potential investment multiplier - banks money as literally

information carriers … carriers of the matter of life investments and thus gaining real competence to create rights on short
for a given system of thinking -> action
scissor bricks Eskm1DAA> m1=EvmDNA

Profits financial investments (describing workloads, competencies, environmental economic efficiency) with the possibility of zero-cost "turnover" with AI robotics
multifaceted trading supporting .. calculation, the physical, biological calculation (environmental -> infrastructural) - but also activities, investments, also as a "marketing" image for, in this case, already a space (environmental) world atomic-microcosmic bank
vectors, control room strategy, multi-plane imaging converging into one measure of project costs and phenomena
hypermacroeconomization of physics, biology, cosmonautics… economics, politics, religion… psychology, pedagogy
economics of science - indicating the directions of consumer investment costs in a macroeconomic approach -> processing, economization of matter ->, i.e., processing -> which is always associated with thriftiness, responsibility, competence-based precisely on thriftiness in its intentions as a guarantee of intentions for the development of existence - rights = carriers - properties of relations in every physical and biological dimension, i.e
competence, personality, thriftiness.. "material" processing room
-> personification of phenomena, processes, matters as economic responsibility -> competence, guarantee of responsibility for non-passive integration with the environment outside and inside in the medical and space science perspective,

including the ageocentric, i.e., a return to … the gods in personification, determination of phenomena.

The cosmic world bank can directly determine the parameters, economy, finances for multiplying the processes of economization of biological, clinical, and physical cosmic phenomena in the selected direction
model m1> m-> physics of biology -> multiplication models of parameters, properties of environmental infrastructure
stakes->stimulation resonance mechanisms

Whatever the structure of the phenomena of matter and life, neo-infrastructural investments in the direction of infinite credits supported by supply (in the infinite perspective of the potential of not matter, but combinations of process efficiency and properties …
fixing the allowance
they give a chance for the development of a comprehensive market economy as well as an increase in the potential of own participation in management, co-responsibility of the competence of the economics of the environment, strengthening its parameters that concern us … parameters …. mass, economic dynamic, process
-> economic control, i.e., an increase in the micro-macrocosm cosmos competence as a material component that determines the direction and speed of development, or degradation in case of cancer or unintentional impact on the climate

The multiplier effect in the combination of infrastructural building blocks of a given structure is more significant with further deeper sub-infrastructural directions of interaction in a given structure.

It is a kind of leverage effect - in the geometrical progression - based on internal, external structures, i.e., infrastructures, changing properties, its potential given our interesting object in common elements resonant with the infrastructural extension, at the lever that can generally increase the potential of the entire object in activities/processes not always directly related to infra/structural changes around the facility.

That is, building an infrastructural advantage inside, outside over, under the environment, to more and more external and internal structures of a given infrastructure, as if over the sub-environment -> operation, algorithmization, transformation at the process level not material, not ... geocentric material, process-flexible, combinational, which have an infinite potential in their most helpful combination, including the combination of energy efficiency, .., mass storage, transport, trans material, ... teleportation in the use of internal infrastructural - media above the structure, which at some point is the structural foundation of a given object of the phenomenon .... further affecting the properties ... durability, resistance ... and vice versa, appropriately manipulating back to weaken these properties, e.g., in cancer ...

m(matter)<m1=p<m2=p1;
m2=EskmDAA m <- p ;
m1=p(process)=EvmDNA or
matter=process(room)  11. C.

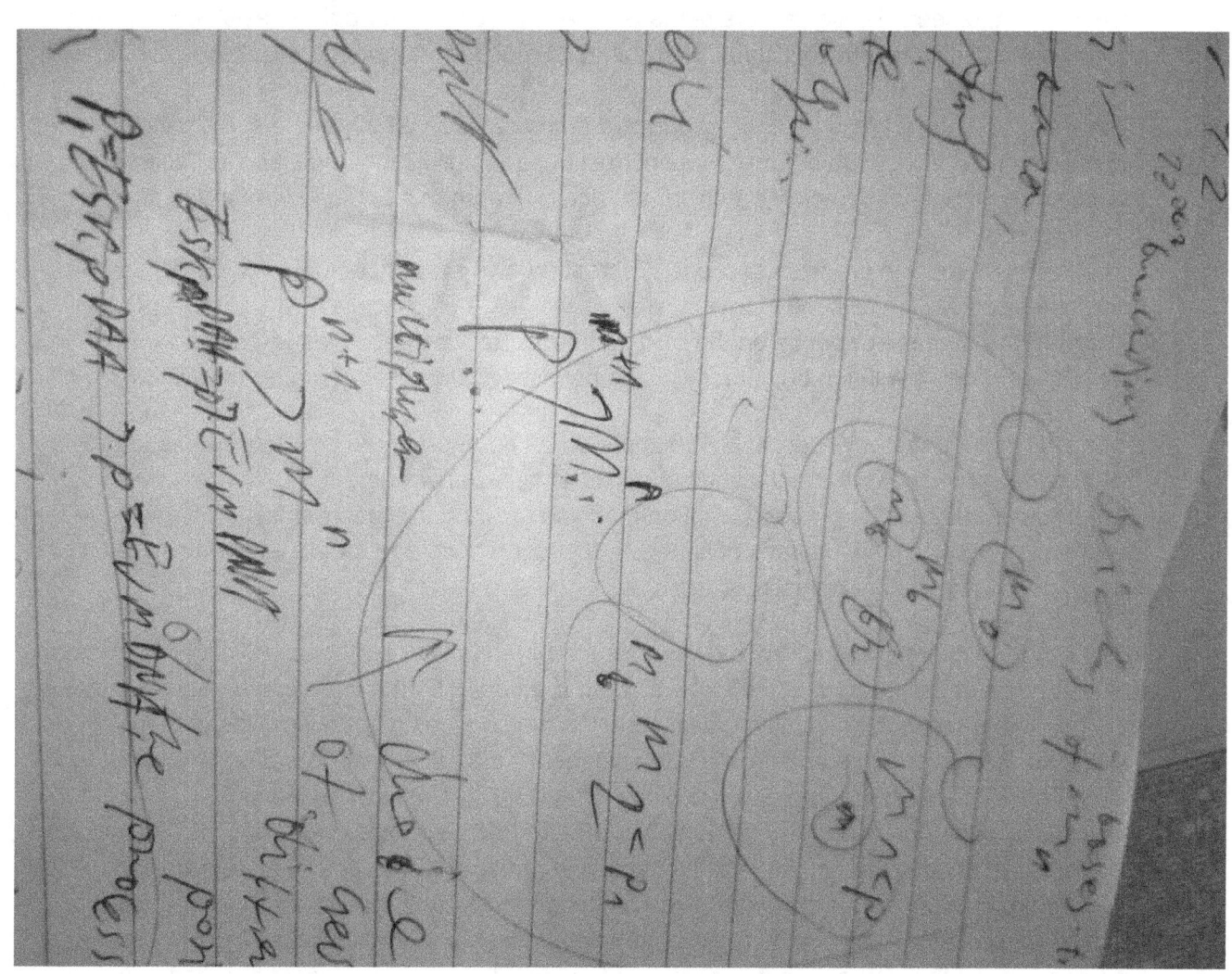

The smallest living->process p building blocks of matter m -micro-macro cosmic theory and practice.

**m<m1=p<m2=p1; m2= Eskm1DAA< Eskm2DAA = p2 ->m<p; m1=p(process)=EvmDNA…**

**p process = m1 processroom > m matter** or breaking all status quo rules of the classic view of matter/life, cosmos, microcosmos, or simply going further beyond current neo geo(ego)centric models.

Further development, suggesting unclosed common resonant, productive = evolutionary infrastructure model development as
procedural status quo of competence m1 and so of the law of m.

Building an internal and external infrastructural advantage over the environment - ecological, algorithmic processes that have an infinite potential of energy, material combinations (low-data phenomena, objects, … life carriers of matter, their properties - durability, resistance, … mass
There are no limits to infrastructural investments in these more and more advanced factors, carriers of matter and life.
There are no natural limits to automation, …AI combinations. Which are to systematically increase their participation in further reconstruction, transformation, evolution, creation of selected fragments or completely new infrastructural fragments of the environment, its absorption…
i.e. directing evolution on given sections, … creating own sections and worlds EvmDNA < p->(m1=EskmDAA) < p1->m2 etc.
The ability to manage depends on the built infrastructural competencies (automation, AI). Directing not only evolution but directly all the necessary elements of matter (and life).

Financial picture strategy-> tax system supported by an appropriate credit system is to provide an accurate picture of the organization of a kind of hyper-environment,
i.e., a neo-infrastructural perpetual motion machine growing in a geometrical progression in the economic support of this process of changing the world once and for all… regardless of other more or less imaginary competitors.

Reducing the interest rate for increasing consumer credits (literally, because this is the meaning of life and its definition), but at the same time strengthened by the classic market, strengthening the tax and infrastructural investments increasing the chances and possibilities of this life, existence, based on constant, growing development - the same it applies directly to the phenomena of processes in physics, biology, chemistry…, classical economics.
And it is all based on one of the fundamental laws of infinite process potential, always ahead of matter.. in every direction of the initiative, action in terms of development or degradation, addicted, on which more or less active, combinational side one stands at a given moment or rather act, process -> always p>m, which pushes, redirects to the successive changes… in exchange for a tax increase that strengthens the infrastructural platform of the market in the classical and environmental terms in us and beyond

, allowing for further increases in these consumer credits and as interchangeable as breathing, etc., simply maintaining life, and more precisely, its material infrastructural carriers, its development, i.e., existence.

So there are no restrictions on infrastructural investments, there are no so-called natural infrastructural barriers, borders because they are of intangible and sub-process potential, ...
with AI automation, they are to systematically stimulate the participation of selected foundations in this reconstruction...... .. the world, i.e., steering on given sections -> the possibility of steering depends on the built infrastructural competences - automation, production .. AI ... is
directs not only evolution but directly all the necessary elements of matter and life
the system of financing taxation and crediting ensures the organizational perpetuum mobile in economic support of this process -> p ...this matter m1...
reducing the rate of increasing the consumption credit - the law of infinite development potential in any chosen direction
-> which pushes the production to the subsequent changes-> in return for a tax increase-> of $100, including a $200 credit ( for infrastructural development of/for more and more responsible(ecology,...)consumption means life!

- in order not to bend the economy, appropriately more extensive credit relief and investment contribution = appropriately compliant directly in physical and biological processes...
  Such public debt, civil, entrepreneurial, and shareholding directly support the consumer market (... warming up to the redness of a given biological, physical relationship - a kind of economic plasma needed for deep material and energy transformations ...
  and reciprocally supporting the consumption market -> increasing infrastructural investments inside and outside us in selected segments through civic participation
- shareholding constant *x push-> one creates the right jobs and ...existence for ourselves and the environment -> a medium that is appropriate up to date.

These neo-infrastructural investments should be strictly related to a given topic, but... they do not have to, and sometimes they cannot - the disadvantages of a kind of incest...
A kind of warming up, leading to the plasma of scientific, economic activity, digging into the back of the environment inside and out, infrastructural development expanding in advance as if without any specific reason and similarity -> these stimulations above the structure is to be a super reaction in the new construction of the cosmos, atom, etc. also in this protection of development, secure economy... carriers of better life and matter, because this is evolution, creation, this is life ... -> not degradation.

DNA in terms of physics or biology with the so-called programmed defects or not, the so-called more or less natural evolutionary barrier that cannot be crossed so easily, so one needs to process more efficiently Esk>Ev p these mDNA organizational and process gaps-> speed up, improve this mechanism of matter development, i.e., carriers physical, biological m... if we are not to collapse so easily (gravity and other terms), or be run over, absorbed by other systems. Such supra-environmental, infrastructural opening of development, consumption, and investment gates.

Furthermore, that is why investments go beyond the framework of the existing systems, relations - old investments being displaced by consumption luxuries - > crediting -> but specific infrastructural investments in ecology, energy, hyper processing - engineering, production ... - matter, life, under the structural process of infrastructural. As if involuntarily pushing out, supporting structures with supermaterial because process-determined EvmDNA =p-> m1<-p EskpDAA ->

undertakings, processes, subprocesses, systems, the world, which are the image of the future matter in an infinite degree of changes in their properties-> processively, where it literally determinates all physical and biological laws, circularly leading to the creation of new internal and external DNA systems, i.e., DAA infrastructure, breaking the contemporary geocentric artificial dogma assumptions according to the so-called contemporary perception and experience are always limited, they do not give proper laws. They are the basis for establishing final laws, which may not affect the level of function, potential but are an old rule of thumb in crossing existential and developmental barriers in relation to other systems inside and outside. It is about more flexible assumptions, available tools, platforms for action, processing, materialization, generation, competence for our existence, proper and position of the micro-macrocosm of the matter of life, i.e. carriers of matter and life, dynamic carriers were p>m.

It is necessary to go beyond the current area, the present structure, at least in our illusions and experiences. To go beyond the substructural one, in order to define its potential with its properties, the potential of its properties, the infinite potential of its properties - and it has to be with maximum momentum. The greater the neo-structural, i.e., infrastructural, momentum of a given object structure, phenomena, undertakings, the greater the chances of the potential, platform, foundation of neo-development, defense of our world ... foundation, neo-process ... the more fantastic the open investment consumption, the greater the infrastructural pressure - breaking bottleneck - above the structural one - more significant neo-generational tendencies -> strengthening a given structure if it is larger not only neutrally existence but also strengthening the existing external-internal environment biological, physical - or rather physical supporting biological, .. hyper ecologically, hyperevolutionary mega-environmental economics planetary engineering hyper economy of space atom.....

# Neoevolutionary Maslow's pyramid of needs, or consumption m < investment bricks p=m1 of developing neo-infrastructure properties of matter... of life, or consumption cannot exceed investment to survive...

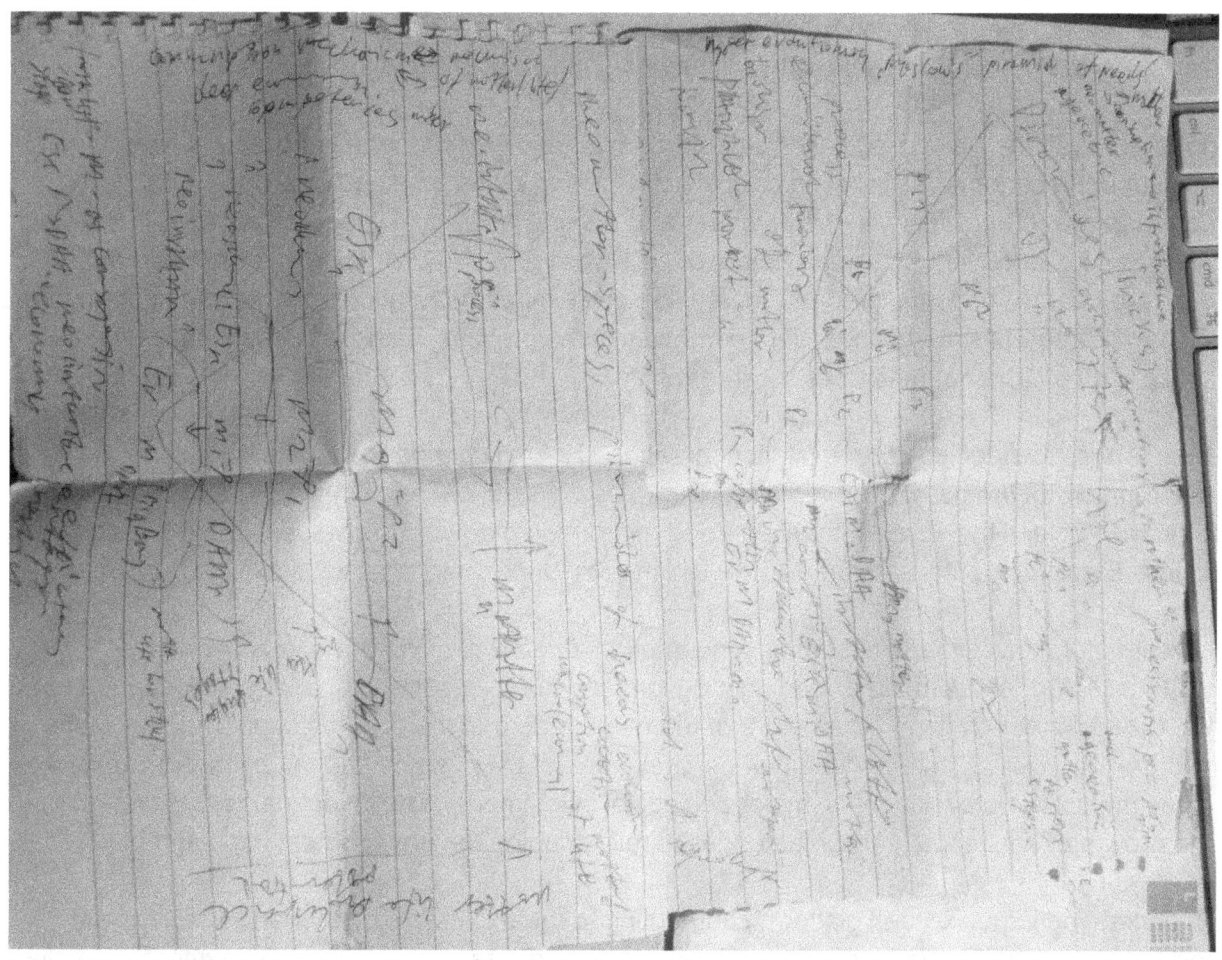

An attempt to bypass the physical, material, involved in the use of the process, economic, even financial, psychological, even artistic, scientific neo-expressionism, dynamic, not passive, expansive, but models, images of competence, neo-evolutionary, creative micro-macro carriers of the cosmos, i.e., man, -> process as processroom, brick in shaping to infinity infinite potential of combinations of bricks, processrooms, which are the basis for expansion or degradation, depending on the consumption or investment advantage as in laying, modeling, building the foundations of any matter, including life, production, technology, ... the artistry of creation - just artistry too!, determination, responsibility, competence based on action, management, including classical (macro) economics, in needs, costs, and own initiative or the initiative of others.

It is simply a hyper anti-geocentric, or rather anti-matercentric vision of shaping, not just a passive attempt to observe the world, adapting to the world, and even more so adapting the world, including ... the entire micro-macro cosmos to our needs, in artistic, economic, i.e., human responsibility, competence supported by hyper evolutionary, hyper ecological, hyperphysical, hyper biological action, process, production, as other systems do diligently.

It is all about streamlining this process as much as possible with tools...automatics, ..AI. so that we can realize dreams, goals, and needs more and more efficiently, which without this proposed model would never have a chance to come true or even lead with further neutral, "sustainable development" - in practice, inactivity, evolutionary indifference, to evident self-destruction.

It requires not the classic, so-called scientific, physical view but the usual consumption, investment, economy, responsible one, always looking at the consequences of exchange, cooperation, competition with the environment, the stake of which is development, without which there is no chance for better parameters of survival in a very dynamic, infinite potential of dangers but also opportunities, i.e., increasing one's infrastructural competencies to determine the laws, properties, and subsequent potentials in the process, in the process of matter, in the process as a carrier of matter, as a carrier of life.

How are we to steer, co-steer evolution, creation? How to change the face of the earth, the micro-macro of the cosmos, in a snap? How to deal with it, because there is no other way of existence unless the abandonment of these micro-macro-cosmic processes - no more matter! unless EvmDNA= processroom m1...n-, which is tantamount to conscious suicide, so... a sin...?

Infrastructural investments, trending consumption, directly reviving the consumer market to whiteness, further liberating it, leading it on infrastructural investments, adequately supported by the tax and credit system, - the nature of the micro-macro of the cosmos supported by macro, micro-economics ... literally -> properties, the shape of the cosmos shaped by direct projects, investments, hyperevolutionary, hyper ecological investments, i.e., investments that underpin the growth of environmental competence, literally shaping the environment in us and beyond.

An increased infrastructural platform contributing to the explosion of the market economy to infinity, supported, framed by further self-propelling process infrastructural investments, literally determining the shape of the matter, or rather the Eskm1...nDAA plasma->environmental process, from the outside accelerating the spindle of the development of the existence of matter as processrooms of carriers life ...understanding its processes and the life of the macro-micro environment of the cosmos.
For higher turnover, degrees of competence, the infrastructure of the pyramid of needs and aspirations, as well as understanding and action, when this pyramid will become a rocket literally and figuratively.

It was about direct management of the environment, i.e., constantly determining costs, credits, ... inflation -> costs -> price value -> but infinite potential constantly shaping this price. Low credit path <- tax, consumptive, but also consumer .... scientific, technological, for infrastructure, for the development of ... scientific competencies in the infinite, bare area of "matter" of process bricks of steps, -> by passively increasing autonomously industrial park, infrastructure park, new environmental potential -> new environment, energy carrier, process > matter, life, as a chance to break through internal and external real and imaginary barriers.

The new infrastructure awakens competencies; it becomes another capsule, a brick in the next stage of development of a given object, a phenomenon, its mastering from the outside and inside.

There is no short quilt effect in public and private investments because, according to the theory of infinity of potential, it is only about stimulating the existing market of needs, changing any combination of bricks - processrooms of infrastructure in their projects.
This market, which pushes the infrastructure by force -> inhale - credit, properly stimulated by exhalation -> expenses, investments (<- taxes, depreciation) infrastructural, or rather neo-infrastructural -> in the 10% income range, for example, -> appropriately reduced interest rate in relation to adequately increased neo-infrastructural taxes/expenditures -> of course, for this there is a need to develop economic tools specifying the mechanism of this neo-environmental neo-infrastructural neo-evolutionism.

This credit factor, or rather the structures of economization of needs, projects, and their elements, can be found in Maslow's inverted pyramid (consumption) of needs...-> consumption $m1$ -> investment $p$ -> consumption $m2$ -> investment $p1$...
It is the infinite potential of pyramidal anti-geocentric - anti-mattercentric exploration.
The issue of making visible directs this mechanism of development of the consumption of "matter" $m$ -> $p=m1$ -> $p1=m2$... for increasing the proper properties of the matter of objects and phenomena in us and beyond, both directly and at a distance.
infrastructure $m < m1$ = inspiration $p < m2$ infrastructure = inspiration $p < m3$ infrastructure .

In order to maintain to exist, one must constantly develop - this is the basis of the existence of any matter, or rather processroom, as a carrier of life, any matter, regardless of its interpretation or naming, application -

That is, **consumption cannot exceed the investment in any application in the economy, physics, biology...to develop/survive** and vice versa if one wants to avoid cancer...

- the so-called life without coverage, without constant, dynamic, non-passive support is literally degeneration, degradation, destruction, suicide on demand, be it in classical economic, physical, biological, and ... moral terms.
That is, an inverted pyramid of needs, where the so-called lower consumption satisfies but additionally supports neo-infrastructural investments, reaching higher stages of production and ... supporting matter

-> process carriers of life in the surrounding environment of competitive, competing systems -> bombarding, pushing, destructively but also supporting, such as wind, gravity in selected, worked out cases of activities.

Constantly investing in EskmnDAA Neo-infrastructural enterprises will allow us to stay afloat.

The greater the neo-infrastructural investments, the greater the chances, exploitation potential, exploration, and transformation of physical and biological relations in the environment; it is also better cooperation in macroeconomics of classical economics, physics, and biology.

Materialcentrism = geocentrism at a given level of the pyramid of projects <-> needs, which can be artificially (infrastructurally) raised in terms of scientific and economic activity, or rather ....eliminated due to new directional investments by newer pyramids of new project needs.

 The new heliocentrism neoinfrastructural pyramid $m<p=m1$ (EvmDNA)$<p1=m2$(EskmDAA)... can give us new views and actions for both economic and scientific issues simultaneously.

www.ingramcontent.com/pod-product-compliance
Lightning Source LLC
Chambersburg PA
CBHW081137170526

45165CB00008B/2705